A TINY PART OF THE REAL UNIVERSE—THE HUBBLE ULTRA-DEEP FIELD. This image, taken by the Hubble Space Telescope, contains about 10,000 very distant galaxies in a tiny area of sky with a diameter only about one-tenth that of the full Moon. In the observable Universe, there are roughly 100,000,000,000 galaxies, each one containing on average roughly 100,000,000,000 stars.
Credit: NASA, ESA, H. Teplitz and M. Rafelski (IPAC/Caltech), A. Koekemoer (STScI), R. Windhorst (ASU), Z. Levay (STScI).

KEPLER

AND THE UNIVERSE

KEPLER

AND THE UNIVERSE

HOW ONE MAN
REVOLUTIONIZED
ASTRONOMY

DAVID K. LOVE

Prometheus Books

59 John Glenn Drive
Amherst, New York 14228

Published 2015 by Prometheus Books

Cover design by Grace M. Conti-Zilsberger
Inset image of Kepler © Corbis
Background image © Dreamstime

Inquiries should be addressed to
Prometheus Books
59 John Glenn Drive
Amherst, New York 14228
VOICE: 716–691–0133
FAX: 716–691–0137
WWW.PROMETHEUSBOOKS.COM

19 18 17 16 15   5 4 3 2 1

Library of Congress Cataloging-in-Publication Data

Love, David, 1950-
    Kepler and the universe : how one man revolutionized astronomy / by David K. Love.
        pages cm
    Maps on endpapers. Includes bibliographical references and index.
    ISBN 978-1-63388-106-8 (hardcover) — ISBN 978-1-63388-107-5 (e-book)
    1. Kepler, Johannes, 1571-1630. 2. Astronomers—Germany—Biography.
3. Astronomy—History—16th century. 4. Astronomy—History—17th century.
5. Brahe, Tycho, 1546-1601. I. Title.

QB36.K4L68 2015
520.92—dc23
[B]

                                                                2015023520

Printed in the United States of America

To Catherine and Rachel

CONTENTS

8   CONTENTS

FOREWORD

The period encompassing the latter part of the sixteenth and early seventeenth centuries heralded what has been termed the "scientific revolution," a crucial component of the European renaissance in thought. And no revolution in thinking could be greater than the changing perception of our place in the Universe. Beginning with the publication of *De Revolutionibus* by Nicolaus Copernicus in 1543, which moved the center of the solar system from the Earth to the Sun, the systematic gathering of the positions of celestial objects soon followed, as well as the development of the astronomical telescope. Despite almost universal opposition in learned circles to these radical concepts, the time was ripe to interpret this emerging data and to physically understand how and why objects move in space.

Within this remarkable period, involving famous observational astronomers such as Tycho Brahe and Galileo Galilei, Johannes Kepler (1571–1630) emerges as the brilliant individual who played this key role of interpretation. In addition to his famous three laws of planetary motion, he was the first to postulate some form of force acting at a distance that governs the positions of celestial bodies. Pre-dating Newton's eventual breakthrough of a theory of gravity by several decades, Kepler's achievements are all the more remarkable given his tragic family circumstances and the bitter opposition he faced throughout his life.

Since Kepler is often placed in the shadow of Galileo, Brahe, and Newton in historical accounts, David Love's excellent and extremely well-researched story rightly restores him to the forefront of arguably the most important scientific revolution of the last millennium. Love describes, with obvious passion, his admiration for this remarkable man and uses his astronomical training to illustrate Kepler's ideas in a manner that is both accessible to a general reader and to the interested specialist. One of the most interesting topics explained in detail is how Kepler defied opposition, bravely adopting Copernicus's concept and demonstrating with impressive rigor how well a Sun-centered solar system matches the celestial motions as compared to an Earth-centered model. Alongside a careful discussion of these great achievements, Love maintains a welcome balance by also highlighting how Kepler made some erroneous deductions and, importantly, places these in the context of contemporary European thought. What is so special in this book is how each of these astronomical developments is interwoven with the full story of Kepler's difficult personal situation. This gives the reader a more rounded view of Kepler and his life.

This is a highly valuable addition to the history of astronomy, complementing more focused and academic treatments because of its broader treatment of Kepler as a person and its lively style, well-informed illustrations, and the clear enthusiasm of the author.

Richard S. Ellis, FRS
Steele Professor of Astronomy, California Institute of Technology
Pasadena, California, June 2015

PREFACE

A long time ago, I found myself reading a lot about the history of astronomy. Ever since, I have been fascinated by two things. First of all, by Johannes Kepler, the extraordinary and brilliant sixteenth- and seventeenth-century astronomer whose life combined a high level of genius and originality with an appalling degree of personal tragedy and suffering. And second, by how and why it is that we have moved from the understandable but woefully inaccurate picture of the Universe that was accepted in Kepler's time (and which Kepler began the long process of replacing) to the radically different insight that we have today.

This book tells Kepler's story within the context of the history of astronomy. He was far more than just one of the first people ever to accept that the Earth went around the Sun. He was a pivotal figure, every bit as important as Copernicus, Galileo, or Newton. It was Kepler who first advocated—in spite of opposition or incredulity—the completely new concept (which today we take for granted) of a physical force, emanating from the Sun, controlling the motion of the planets in their orbits. And the three laws of planetary motion that he discovered as a consequence describe (to a high degree of accuracy) the way that the planets and their moons move.

It was Copernicus who introduced the idea of a Sun-centered system to the medieval world. But it was Kepler who went on to show beyond any reasonable doubt that this system worked far better than the old Earth-centered view. His work marked a huge advance in the physical sciences, and he is rightly thought of as one of the key founders of the scientific revolution.

His huge productivity is all the more surprising when considered against the background of the dreadful amount of suffering in his life. He had an unhappy childhood. He suffered from health problems throughout his life. He fathered twelve children, eight of whom died in infancy or childhood. His first wife and his much-loved stepdaughter also died at relatively young ages. His mother was accused of being a witch and narrowly escaped the death penalty. He was also frequently caught in the three-way dogfight between Lutherans, Catholics, and Calvinists that was the defining feature of the age. Yet in spite of it all, and after a shaky start, he acquired a large and loyal following of friends and admirers, often across the religious divides of the time.

It should not be assumed that Kepler always got everything right—far from it. Given the benefit of our modern understanding of the Universe, we can see that he was often wrong. Sometimes this was because he was simply reflecting the worldview that was prevalent at the time, and sometimes this was a function of Kepler's own unique approach.

Yet paradoxically, his first and biggest wrong idea—the connection he incorrectly thought he saw between certain geometrical shapes and the orbits of the planets, and which he believed he had found as a result of divine inspiration—ended up bearing fruit in other areas. His idea provided the motivation to continue in his work (which eventually led to his discovery of his three laws of planetary motion), but it was in fact completely wrong.

It is the combination of a genuinely likeable personality with a life of tragedy, and the contrast between a deep insight into the nature of reality and a hopelessly wrong mysticism, that makes Kepler such an endearing and fascinating character in the story of the scientific revolution.

Kepler cannot be understood in isolation. So the introduction to the book gives a brief history of astronomy from the ancient Greeks up to his own time. Equally, our understanding of the Universe since Kepler has changed fundamentally and in ways that he could never have guessed (and would certainly have been dismayed by). To illustrate the vast gulf in scientific awareness between Kepler's time and ours, the final chapter summarizes some of the major astronomical advances since his time and concludes with a brief explanation of our current understanding of the origin of the Universe.

In writing this book, I have gone back to as many of the major primary source documents as possible, and I am grateful to a number of publishers and authors for permission to use quotations from their translations. These are all listed in the acknowledgments at the end of the book. A complete list of the numerous people who have helped me would take up too much space. However, I wish to mention in particular Professors Richard Ellis and Ian Morison, for having read through various chapters in the book and having offered some helpful suggestions; Professor Owen Gingerich, for confirming that my understanding of certain key dates was correct; the late Peter Hingley, the Royal Astronomical Society librarian, for his helpfulness at all times; and Mario Micciché, for having translated a large number of Kepler's key autobiographical writings from Latin for me. I am also grateful to George Wilkins and other members of the History of Science group at the Norman Lockyer Observatory Society for the numerous useful discussions I have had with them over the years.

I would also particularly like to thank Claudia Hehmann, Hermann Faber, and Wolfgang Schütz, at the Kepler Museum in Weil der Stadt, all of whom provided friendly and enthusiastic help and willingly answered a number of questions that I had. Lothar Sigloch, the head of the Stadtarchiv in Weil der Stadt, kindly allowed me to look through a number of volumes of *Johannes Kepler Gesammelte Werke* that were not easily obtainable in Britain.

I am also very grateful both to Euratlas Maps for the excellent software that allowed me to create three maps of parts of Europe as they were in the year 1600, and to SkyMap Pro for the equally excellent software that allowed me to create two star maps.

My penultimate thanks must go to Steven L. Mitchell, Pete Lukasiewicz, Mary Read, Cate Roberts-Abel, Julia DeGraf, and others at Prometheus Books for having expertly guided me through the mysteries of the publishing process.

Finally, I am grateful to my wife, Dr. Elizabeth Emerson, for accompanying me on my travels around Europe to visit the places where Kepler lived and worked, and for her suggestions and comments, and also to both her and our daughters, Catherine and Rachel, for their constant support and encouragement, and to Rachel for her translation skills. Responsibility for any remaining errors in the text is entirely my own.

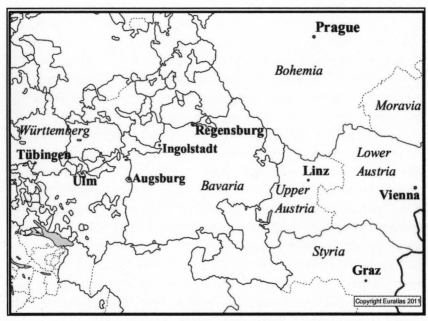

Part of the Holy Roman Empire in 1600
(now southern Germany, Austria, and the Czech Republic)

ASTRONOMY BEFORE KEPLER

THE ANCIENT GREEKS: PYTHAGORAS

The story of astronomy begins with the ancient Greeks. Others before them had studied the stars and planets and had observed their positions and regularities. But it was the ancient Greeks who made the first serious effort to provide a physical explanation (albeit incorrect) of what was happening in the heavens.

One of the most important figures in this quest was Pythagoras, who was born on the Greek island of Samos around 570 BCE, then settled in southern Italy, where he gathered a number of followers around him, and who died in about 500 BCE. He was believed by some to be the son of a merchant called Mnesarchos, and by others to be the son of the god Apollo.[1] Like Kepler, he was both a mathematician and a mystic, and he and his followers have had a huge influence on later thought. Unfortunately, we only know about him and his ideas from later writers.[2]

However, it seems likely that Pythagoras was the first person to realize that the Earth was a sphere. He (or more probably just his later followers) also believed that the Earth was in motion, but not around the Sun. Instead, it was thought that the Earth revolved around a central fire, always keeping the same hemisphere pointing toward this central fire (just as the Moon always keeps the same face pointing toward the Earth). Greece and Italy were both located on the far side of the Earth, so the central fire could never be seen from these locations. (In the same way, somebody living on the far side of the Moon would never see the Earth.)

Pythagoras also realized that everything in nature was governed by numerical relationships. This was a vital insight that was crucial to the later development of science. He is meant to have originated the saying that "all things are numbers." Regularity showed itself, for example, in music—harmonious sounds could be produced by plucked strings whose lengths were in simple numerical relationships with each other. The planets themselves, he argued, all produced musical notes that were in harmony with each other—the music of the spheres. Later legend maintained that Pythagoras was the only one who could actually hear these planetary notes. He is also famous, of course, for the theorem named after him: that the square of the longest side in a right-angled triangle is equal to the sum of the squares of the other two sides.[3]

Two thousand years later, Kepler was to remark that it was Pythagoras who had first understood the importance of certain geometrical shapes that Kepler was to make great use of. Kepler made frequent mention of Pythagoras in his writings, speaking of him as "the grandfather of all Copernicans"[4] and referring to him and Plato as "our true masters."[5]

After it was accepted that the Earth was a sphere, another ancient Greek, Eratosthenes, who lived in the third century BCE, managed the astonishing feat of measuring the size of the Earth. Eratosthenes was the head of the famous library at Alexandria, in Egypt. He knew that some way south, at Syene (on the site of modern-day Aswan), it was possible to see the Sun directly overhead from the bottom of a deep well at midday in midsummer. In Alexandria, the Sun never quite rose this high in the sky. Eratosthenes realized that if he measured the angle by which the Sun's maximum height deviated from the vertical at Alexandria, and if he knew the distance from Alexandria to Syene, it was a matter of simple geometry to calculate the circumference of the Earth. Although there is some uncertainty about the size of the units he used, it does seem that the figure he arrived at was within a few percent of the modern value.[6]

PLATO AND ARISTOTLE

As far as the idea of a moving Earth was concerned, astronomy took a small step backward with Plato (428–348 BCE). Plato still recognized that the Earth was a sphere (as have most educated people in the Western world ever since[7]), but he did not think it moved; instead, he thought it was fixed at the center of the Universe. The Earth—he believed—was surrounded by, and at the center of, a vastly bigger sphere (the celestial sphere), to which all the fixed stars were attached. The celestial sphere revolved around the Earth from east to west in about twenty-four hours, which explained why the fixed stars drifted across the sky as the night went by.

But in addition to the fixed stars, there were seven wandering bodies that were clearly not attached to the celestial sphere. These seven bodies, the Sun, the Moon, Mercury, Venus, Mars, Jupiter, and Saturn were constantly changing their positions relative to the stars on the celestial sphere.[8] (The planets Uranus and Neptune would not be discovered until more than two thousand years later.) These wanderings marked them as being quite different from the stars. Indeed, the modern word *planet* is derived from the ancient Greek word for "wanderer."

Plato envisaged a system in which the key principle was that of uniform circular motion. The Sun, the Moon, the planets, and the stars should ideally move in perfect circles at uniform speed, carried around the Earth by gigantic, transparent, and weightless crystalline spheres. His pupil Aristotle (384–322 BCE) thought the celestial bodies were composed of a fifth element—a quintessence—that was both superior to and quite different in nature from the elements on Earth, and incapable of change. For this substance, circular motion around the Earth was the natural order. This was in contrast to the elements of earth and water (which naturally moved toward the center of the Earth) and air and fire (which naturally moved away from it).[9]

So ideally, the Greeks would have liked to see planetary motion explained by the system shown in figure I.1. The Sun certainly followed this pattern fairly closely (fig. I.2). In addition to the Sun's daily motion around the Earth, it was possible to work out its changing position during the year relative to the background of fixed stars by noting the positions of the stars just after sunset or just before sunrise. It was clear that the Sun moved along the same fixed path (called the ecliptic) relative to the stars, at an almost constant rate, returning to essentially the same spot relative to the stars after one year, and that it did this year after year.

However, the planets refused to behave like the Sun. They all moved in paths very close to the ecliptic, but certainly did not do so at a steady rate. Take Mars as an example. Figure I.3 shows how it behaves. Initially, it moves from west to east (relative to the background of fixed stars), at something close to a constant rate most of the time. But then it slows down and reverses its direction for a few months, before resuming its original course. The other planets also display this peculiar retrograde behavior.

Why should the planets behave in this strange way? That was the central mystery.

The Greeks set themselves the task of working out how it could be explained, while still retaining the all-important concept of circular motion at uniform speed. The solution they finally settled on was advocated by Apollonius of Perga, a Greek mathematician who lived in the third century BCE.

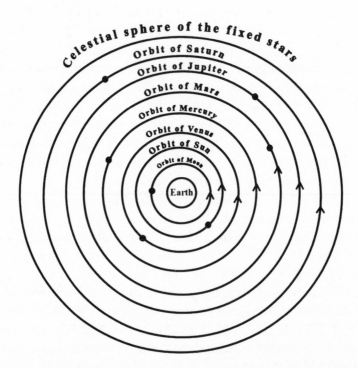

Figure I.1. Plato's ideal system, in which the planets, the Sun, and the Moon move around the Earth in perfect circles at uniform speed. Later Greek philosophers interchanged the orbits of Mercury and the Sun.

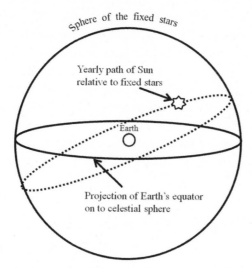

Figure I.2. The Sun's yearly path relative to the stars—the ecliptic. (In reality, of course, it is the Earth's movement around the Sun that causes this apparent motion, but this would not be generally realized for another 2,000 years.)

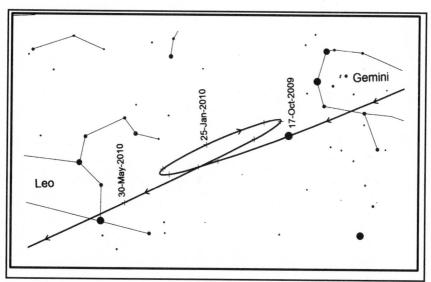

Figure I.3. The apparent "forward-backward-forward" motion of Mars during late 2009 and early 2010. Initially, Mars moved from right to left, through the constellation of Gemini and on toward Leo. From late December until mid-March it reversed this direction, then resumed its original course. (SkyMap Pro software.)

Apollonius proposed that the planets were not directly orbiting the Earth, but that each planet was traveling around an epicycle, a small circle whose center revolved in turn around the Earth in a much bigger circle (called a deferent), as shown in figure I.4. The effect of the epicycle was temporarily to reverse the normal direction of movement of the planet in the sky. By adjusting the size of the epicycle (relative to the deferent) and the speed the planet traveled around the epicycle, the Greeks found that they could come up with a good first approximation of each planet's strange movements.[10]

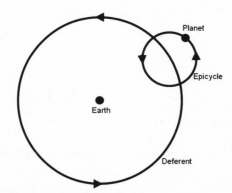

Figure I.4. According to Apollonius, each planet traveled around an epicycle, and it was the center of the epicycle, not the planet itself, which revolved around the Earth.

In one of the ironies of history, Apollonius was also the first person to discover conic sections—slices through a cone that (depending on the angle at which the slice is made) produce in turn the circle, the ellipse, the parabola, and the hyperbola. But it would have to wait some two thousand years before Kepler was to discover that the planets move in ellipses rather than circles.

PTOLEMY

A later Greek astronomer, Ptolemy (90–168 CE), recognized that work was still needed on the system. The observations showed that the center of each planet's epicycle moved more quickly on one side of its (assumed circular) orbit around the Earth than the other side. To allow for this, he tried adjusting the position of the Earth so that it was offset a little from the center. (He reasoned that if the planet really was moving in a circular orbit at constant speed, then offsetting the Earth from the center would give the observed appearance of variable speed.)[11]

But if he adjusted the Earth's position far enough to create the required apparent variation in speed, it implied that the size of the epicycle must be varying. So instead he introduced a further fudge factor—the equant. The equant was an ingenious idea. It was a point in space that was also offset from the center, but in the opposite direction from the Earth and from which the center of the epicycle merely *appeared* to move at a uniform rate.

By suitable adjustment of these two fudge factors, the offset Earth and the equant (both of them departures from the original Greek ideal), Ptolemy found that he could replicate the apparent motions of the planets tolerably well. The best fit came if he assumed that the Earth and the equant were equally distant from (and on opposite sides of) the center (fig. I.5).

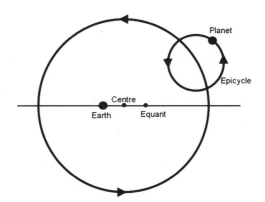

Figure I.5. To try to align theory and observation, Ptolemy introduced two fudge factors, the offset Earth and the equant, both of them departures from the original Greek ideal.

However, the concept of the equant was a clear divergence from the principle of circular motion at a uniform rate.

From the vantage point of the equant, the center of the planet's epicycle only *appeared* to move uniformly. In reality (according to Ptolemy) its speed increased and decreased over its orbit. Although this seemed to be the only credible way of explaining the observations, it was regarded by many later astronomers as a fudge too far and was not liked. (What nobody realized, and what was to be one of Kepler's most important discoveries, was that the hated equant was actually saying something important about the reason for the planets' strange behavior.)

The elaborate and contrived nature of the Ptolemaic system also led many to conclude that it was simply a means of calculating future planetary positions rather than a true representation of reality, which was perhaps impossible to attain.

After Ptolemy, Greek astronomy came to an end. The Greek texts were preserved by the Arabs and the Persians, who also refined some of Ptolemy's techniques but failed to make the only breakthrough that mattered, and left astronomy much as they had found it. (They did, however, manage to determine certain astronomical constants very accurately—for example, the brilliant poet and mathematician Omar Khayyám measured the length of the year to a higher level of accuracy than was achieved in the West for many centuries. They also gave names to many stars that are still in use today—Aldebaran, Algol, Betelgeuse, and several others.)[12]

The West, meanwhile, languished in scientific dark ages. Awareness of Greek ideas was lost for many centuries, and it was only the gradual recovery of the ancient Greek texts (from Arabic sources) that slowly brought about a revival of knowledge. The consequence was that nothing of any great significance happened in Western astronomy for the next fourteen hundred years.

PROBLEMS WITH PTOLEMY

By the standards of the time, the Ptolemaic system gave tolerably good results. But with the benefit of hindsight, it is easy to see that (if it were to be regarded as physically real rather than merely as a calculating device) it was full of problems. Why was it, for example, that Venus and Mercury both appeared to take the same time (on average) as the Sun to go once around the Earth, even though they were meant to be in independent orbits? Why was it that whenever Mars, Jupiter, or Saturn was in the middle of its strange retrograde motion, the Sun was always to be found exactly 180° away, on the opposite side of the celestial sphere? In short, why did the Sun seem to exercise such control over the planets?

Why also was the system just so complicated? It was this sheer complexity that gave rise to the famous (and perhaps apocryphal) statement by Alphonso X, the thirteenth-century king of Castile, that "if God Almighty had consulted *me* before embarking on creation, I would have recommended something a little simpler."[13]

NICOLAUS COPERNICUS

By the time of Copernicus (1473–1543), the Ptolemaic system was still thought to be broadly correct. The West's view of the Universe at that time is perfectly summarized by the illustration shown at the front of this book. This is from Hartmann Schedel's *Weltchronik*[14] (usually known as the *Nuremberg Chronicle*), a popular book published in 1493 and the historical *Wikipedia* of its day. (It has been suggested that some or all of this illustration may have been designed by German painter Albrecht Dürer, who was born exactly one hundred years before Kepler.) The Earth was shown at the center of the Universe, surrounded by the other elements of water, air, and fire. Beyond the Earth were the Sun, the Moon, and the five known planets, all in motion around the Earth. Beyond these lay the sphere of the fixed stars, and beyond this outermost sphere were God and Heaven. The Universe was finite in extent, and humans (for whom it seemed obvious that the Universe was made) were at the very center of it.

Copernicus was born in Toruń, in modern-day northern Poland. His original surname was Koppernigk, but in his sixties he Latinized it to the more dignified form of Copernicus.[15] His father died when he was young, and his maternal uncle took over responsibility for his education and welfare. Uncle Lucas Watzenrode was a bishop in the Roman Catholic Church and found his nephew a job for life as a canon at Frombork cathedral, in the north of Poland. This blatant act of nepotism—a not unknown practice in the Catholic Church of the time[16]—turned out to be a very good thing for the future of science. Even though the post of canon was a busy one, it nevertheless gave Copernicus the time and the security he needed to be able to devote much of his life to astronomy.

It is sometimes wrongly stated that Copernicus was the first person ever to suggest that the Earth went around the Sun. In fact, he was not. Probably the first person ever to do so was Aristarchus of Samos, an ancient Greek who lived in the third century BCE, although his idea was not taken seriously.[17] Copernicus seems to have been dimly aware that this may have been Aristarchus's view because it receives a brief mention in the first draft of his great book (although—mysteriously— not in the final version). However, the difference between Copernicus and Aristarchus is that Copernicus didn't merely make the suggestion that the Earth and the other planets went around the Sun. He developed a complete geometrical model based on this assumption. Anybody else in the previous fourteen hundred years could have come up with the ideas behind his system—all the evidence was there—but nobody did. In short, he was a genius and is justly famous for his work.

He adopted the revolutionary idea of a Sun-centered Universe in spite of the fact that the Ptolemaic system was not broken and didn't need fixing. Admittedly it gave far from perfect forecasts of future planetary positions. But the Ptolemaic system was in accord with common sense. In reality, of course, as well as its rotation on its own axis every twenty-four hours, the Earth is moving around the Sun at about 30 kilometers (about 19 miles) every second. It also shares in the Sun's motion around the center of our Galaxy at about 230 kilometers (about 140 miles) every second. To us, though, the Earth seems to be completely motionless, and—to the unaided eye—all the heavenly bodies seem to be moving around us. The Ptolemaic system also lined up well with the teaching in the Bible that the Earth was fixed and did not move. (The biblical texts were later to prove an important stumbling block in the way of acceptance of change.) So the key to understanding Copernicus's reasoning lies in the fact that a Sun-centered system has a certain simplicity and internal logic that is completely absent from the Ptolemaic system.

Copernicus had realized that there were—in principle—two enormous advantages of a Sun-centered Universe over an Earth-centered Universe. First of all, he saw that he could automatically account for the strange retrograde motion of the planets as an inevitable consequence of observing one moving planet from another moving planet.[18] Figure I.6 explains why. As Mars moves around the Sun, from position 1 to position 7, the Earth also moves around the Sun. If the position of Mars (as seen from a moving Earth) is projected on to the background of the fixed stars, it *appears* to move forward, then backward, then forward again. The apparent temporary backward motion arises as a direct consequence of observing one moving planet from another moving planet, and is entirely illusory. The apparent backward motion is just a reflection of the faster forward motion of the Earth in the part of its orbit where it overtakes Mars.

It is rather like a faster train overtaking a slower train. Passengers in the faster train can sometimes get the false impression that the slower train is actually moving backward. For the previous two thousand years, people had been fooled by much the same optical illusion into believing that planetary motion was far more complex than it actually is.

Copernicus had moved away from the essentially descriptive (and unnecessarily complex) approach of Ptolemy to what the planets seemed to be doing, and toward a simple underlying explanation of their strange behavior. The apparent retrograde motion of each planet was merely a reflection of the Earth's own motion in its orbit around the Sun. The five different epicycles, one for each of the other planets, could be replaced by the motion of a single planet—the Earth.[19]

There was a second advantage. In the Ptolemaic system, it was not possible to determine the distances to the planets. The only distance that could be measured was that of the Moon, which is easily the closest body to Earth, and which was known to be roughly 30 Earth diameters away. Beyond that, planetary distances were little better than guesswork. However, once it is realized that the Earth is a moving platform, this provides a means of measuring relative (but not absolute) distances.

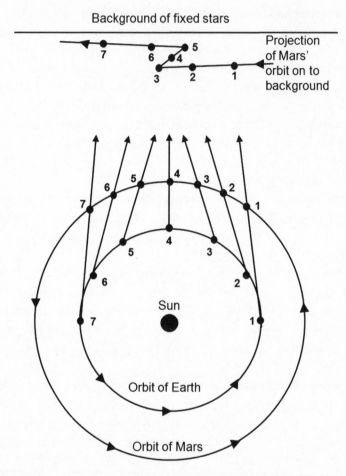

Figure I.6. Apparent retrograde motion is the inevitable result of observing one moving planet from another moving planet.

Copernicus found that, on the assumption of a Sun-centered system, he could calculate the time it took each planet to go once around the Sun. He could also calculate each planet's relative distance from the Sun (taking the Earth–Sun distance to be equal to 1). He realized that the two lists lined up with each other—the closest planet to the Sun (Mercury) circled it in the shortest period of time, and the farthest known planet from the Sun (Saturn) took the longest time to go once around. And the Earth very neatly fell into position three in each list:

Table I.1. Planetary periods and relative distances from the Sun.

	Relative distance from Sun (modern values)	Time to go once around Sun (modern values)
Mercury	0.4	88 days
Venus	0.7	225 days
Earth	1.0	365 days
Mars	1.5	687 days
Jupiter	5.2	11.9 years
Saturn	9.5	29.5 years

In the Ptolemaic system, there had simply been no way of deciding the order of the Sun and the planets Mercury and Venus in terms of their relative distances from a central Earth. Different ancient Greeks had assigned different orders to them. In contrast, in Copernicus's system, determining the order of distances of planets from the Sun had become a straightforward series of calculations. There was a compelling internal logic to the new system.

One of the few passages in Copernicus's book where he is clearly trying to sell his new idea brings out this distance-time relationship: "Therefore in this ordering we find that the world has a wonderful commensurability and that there is a sure bond of harmony for the movement and magnitude of the orbital circles such as cannot be found in any other way."[20]

From a modern perspective, there can be no doubt that Copernicus came up with what in principle was a simpler system than Ptolemy, with a much greater explanatory value. At the time, though, things were not so clear-cut. The embarrassing thing about the Copernican system is that in practice its predictions were hardly better, if at all, than those of the Ptolemaic system. This was partly because Copernicus never thought of moving away from a system dominated by uniform circular motion. (It took the genius of Kepler, sixty years later, to do this.) And one of Copernicus's principal motivations was to get rid of the much-disliked equant, with its concept of variable speeds. He did this by introducing an alternative geometrical method of his own,[21] and by doing this he arguably made his system in some respects even more complicated than Ptolemy's.

Some have perversely tried to argue that Copernicus's achievement was not such a huge turning point in history. It is true that Copernicus would never have achieved what he did if the works of the ancient Greeks had been lost or had remained untranslated. It is true that much of his thinking was dependent on the ancient Greeks, and on Ptolemy in particular. It may well be true that some of his arithmetical methods were derived from Arabic sources. But all this pales into utter insignificance beside the colossal revolution in thought needed to move the Earth away from the center of the Universe. Copernicus achieved a massive conceptual breakthrough. He himself never knew it, but he truly was one of the founders of what can justifiably be called the scientific revolution.

Copernicus kept the details of his idea hidden (by his own account) for about thirty-six years,[22] although he circulated an early initial account to friends.[23] But he might never have published his ideas had it not been for Georg Joachim Rheticus, a professor of mathematics from the Lutheran stronghold of Wittenberg University. Rheticus, alone among his contemporaries, became convinced of the truth of the Copernican system, traveled to Frombork and spent about two years with Copernicus, both getting to grips with the detail of his system and trying to persuade him to publish it. Copernicus was reluctant. Why did he delay publication for such a long time? The only clue we have lies in a single piece of correspondence: Andreas Osiander, who will enter the story later, wrote to Copernicus about "the peripatetics [i.e., the Aristotelians] and the theologians whom you fear will raise objections."[24]

Eventually, the elderly Copernicus was persuaded to publish. But his health was failing. Rheticus took the manuscript to Nuremberg and set in motion the printing process, but he left for another university post in Leipzig (over 200 kilometers [about 140 miles] away) before it was complete. Copernicus's health grew worse, and he suffered at least one major stroke. Unable to leave his bed, he finally received a printed copy of his book, *De Revolutionibus Orbium Coelestium* (*On the Revolutions of Heavenly Spheres*), on the afternoon of May 24, 1543. By the evening, he was dead.[25]

He was buried in Frombork cathedral, as would be appropriate for a canon. What is curious is that his grave was unmarked for several centuries. The remains of his body were only eventually identified following their exhumation in 2005.

Apart from the first section, his book is an impossibly difficult read for a layman. It has only a few passages of inspired writing, one of which is this passage on the role of the Sun: "In the center of all rests the Sun. For who would place this lamp of a very beautiful temple in another or better place than this, wherefrom it can illuminate everything at the same time? As a matter of fact, not unhappily do some call it the lantern; others the mind, and still others the pilot of the world. . . . And so the Sun, as if resting on a kingly throne, governs the family of stars which wheel around."[26]

In fact, the center of all movement in the Copernican model was the center of the Earth's orbit, not the position of the Sun (although the two were not very different). Nevertheless, it is possible that it was this hymn of praise to the Sun that stimulated a young Johannes Kepler to give the Sun a physical control over the planets that had never been thought of or attempted before.

REACTIONS: MARTIN LUTHER

There were a number of different reactions to Copernicus's ideas. In typically forthright tones, Martin Luther (1483–1546), one of the founders of Protestantism and a contemporary of Copernicus, thundered that "this fool wishes to reverse the entire science of astronomy; but sacred scripture tells us that Joshua commanded the Sun to stand still, and not the Earth."[27] The Book of Joshua, part of the Old Testament, tells the story of how the Israelites moved from the wilderness to the Promised Land. Before they could occupy their new land, they first had to kill the tens of thousands of men, women, and children who were already living there. To assist in the carnage, by providing several extra hours of valuable daylight, Joshua commanded the Sun (not the Earth) to stand still (Josh.10:12).

To the modern mind, the curious thing about this story is that Martin Luther never for a moment questioned the morality of what the Israelites were doing. Instead, he simply found this convincing evidence that it was the Sun that moved and the Earth that was fixed.[28]

Nor was he alone. Even the mild-mannered Philip Melanchthon (1497–1560), the leading intellectual of the Lutheran movement, whose studies at Tübingen University had included astronomy, said that "certain men, either from the love of novelty or to make a display of ingenuity, have concluded that the Earth moves. ... Now it is a want of honesty and decency to assert such notions publicly, and the example is pernicious. It is the part of a good mind to accept the truth as revealed by God and to acquiesce in it. ... The Earth can be nowhere if not in the center of the Universe."[29] Like others, Melanchthon seems to have admired Copernicus's mathematical techniques, but he never changed his mind about the Earth's central position.[30]

Other Protestant leaders held the same view, based on a number of biblical passages,[31] such as the words of Psalm 93: "Thou hast fixed the Earth, immovable and firm." John Calvin (1509–1564) was in no doubt that the Earth was at the center of the Universe. The remark attributed to him: "Who will venture to place the authority of Copernicus above that of the Holy Spirit?" is probably apocryphal,[32] but it is certainly the sort of thing he would have said if he had known about Copernicus's ideas.

The reaction of the Protestants was hardly surprising. They had taken their stand against Catholicism on the basis that what was needed was a return to the teachings of the Bible. They could hardly now concede that certain passages of the Bible were not literally true.

ANDREAS OSIANDER

A more nuanced approach came from Andreas Osiander (1498–1552), another Lutheran clergyman. After Rheticus had left for Leipzig, Osiander ended up with the responsibility for printing Copernicus's book. Unknown to the author (who was by then lying on his deathbed), he inserted an unsigned preface into the book. The preface said that the proposed Sun-centered system was not to be regarded as true but was "only in order that they may provide a correct basis for calculation."[33] This was not simply a reflection of Osiander's desire to shield the book from criticism. It also reflected the widely held view (known as instrumentalism) that different astronomical theories were not about truth; they were all simply calculating devices. This view had arisen partly because of the sheer complexity of Ptolemy's system, and partly because it was well established that two apparently different mathematical models of planetary motion gave identical results.[34] The one and only certainty was the Earth's central position in the Universe. Beyond that, mathematical models of planetary motion said nothing about physical reality, which was widely believed to be unknowable.

So for many decades after, most people took the view that the Copernican system was no more than a useful calculating device—an alternative and conceptually superior way of forecasting planetary positions but certainly not a true statement of the way things actually were. It was also widely believed (until Kepler made the truth of the matter public[35]) that Copernicus himself, not Osiander, had written the preface. Copernicus would have strongly disapproved of the preface if he had known about it, and it is clear from the main text of his book that he believed in the physical reality of his theory. However, it may well have been the preface that postponed the banning of the book by the Catholic Church for the next seventy years.[36]

TYCHO BRAHE

A further approach came from Tycho Brahe (1546–1601), the best observational astronomer not just of his own time but in the whole history of pre-telescopic observational astronomy. Tycho was an important figure in Kepler's story. He was a rich Danish nobleman who devoted his life, and his not inconsiderable resources, to determining far more accurate planetary and stellar positions than had ever been achieved before. He did not simply make better observations of planetary positions—critically, he also made regular observations of planets at numerous positions in their orbits.

Tycho was hostile to the Copernican system on the solidly observational basis that, if it were true, we would expect to detect some stellar parallax. Parallax is the phenomenon by which a nearby object seems to move (normally against a background of more distant objects) as an observer changes his position. If the Earth really did move around the Sun, Tycho reasoned, then (unless the celestial sphere was unimaginably distant, which did not seem credible) stars on the celestial sphere would be seen from different angles at different times of the year and would change their apparent positions very slightly. Tycho had made the most accurate measurements ever of stellar positions. His measurements were accurate to about two minutes of arc (2/60 of a degree). Yet he saw no sign whatsoever of any parallax.

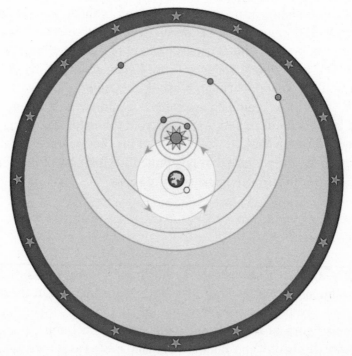

Figure I.7. Tycho's compromise system, which preserved a stationary and central Earth and a Sun that moved around the Earth but made the five planets revolve around the Sun.[37]

Nowadays, measuring stellar parallax is the way we have calculated fairly accurate distances to over one hundred thousand of the closest stars. But even Proxima Centauri, the nearest star, has a parallax of less than one second of arc (i.e., less than one-sixtieth of one-sixtieth of a degree), which is smaller than one-hundredth of the most accurate measurement that Tycho could make. (Proxima Centauri is actually at a distance of about 40,000,000,000,000 kilometers.) So the stars are vastly farther away than Tycho (or almost anybody else at that time) had imagined, and it is not at all surprising that Tycho couldn't detect any stellar parallax.

Tycho concluded that the Earth didn't move around the Sun. He constructed a compromise system that was halfway between Ptolemy and Copernicus, and preserved a stationary Earth at the center of the Universe. The Sun (and the Moon) continued to orbit around the Earth, but all five planets moved around the Sun, as shown in figure I.7. This ingenious compromise on the one hand maintained the stationary Earth but on the other hand gave many of the advantages of simplicity and internal logic of the Copernican system. It attracted many followers.

ERASMUS REINHOLD

Even though Copernicus's proposed cosmology was largely ignored or attacked, his mathematical methods were widely accepted not merely as valid but as an improvement on existing ones. This was partly because they eliminated the need for the much disliked equant. Erasmus Reinhold (1511–1553) was the professor of mathematics and astronomy at Wittenberg University. In 1551, he used Copernicus's methods to produce a set of tables—the *Prutenic Tables*[38]—from which planetary positions could be predicted. These tables were a significant improvement on, and gradually replaced, the *Alphonsine Tables* (named after King Alphonso X), which had been in use for the previous three hundred years.[39] They also had the effect of making Copernicus's work better known. However, they scrupulously ignored the fact that Copernicus's calculations had been based on his Sun-centered theory.[40]

MICHAEL MAESTLIN

A more positive reaction came from Michael Maestlin (1550–1631), who was a professor of mathematics at Tübingen University.[41] Maestlin, although deeply reluctant to say so publicly, was one of the very few people in the whole of Europe who recognized that the Copernican system was literally true. (Thomas Digges and William Gilbert[42] from England and Giordano Bruno from Italy were three others.[43]) He was another important figure in Kepler's story, and we shall be returning to him later.

Maestlin was in a small minority. People admired Copernicus for his mathematical skill in getting rid of the equant, and because his work provided the basis for the *Prutenic Tables*. However, the fact of the matter was that the Ptolemaic system gave tolerably good predictions of planetary positions by the standards of the time. And plain common sense, Aristotelian physics, the lack of any detectable stellar parallax, and the teaching of the Bible all combined to demonstrate the utter absurdity of the idea that the Earth moved.

CHAPTER 1
KEPLER'S EARLY LIFE

The year was 1571. Copernicus had been dead for the previous twenty-eight years, and his idea of a Sun-centered Universe had still received virtually no public support. Tycho Brahe was a young man of twenty-five. William Shakespeare and Galileo Galilei were both seven years old. Queen Elizabeth I of England had been on the throne for thirteen years. William Gilbert was just arriving in London to practice medicine. And Johannes Kepler was born on Thursday, December 27, in the Free Imperial City of Weil der Stadt, in southern Germany, the first child of Heinrich and Katharina Kepler.[1] A later autobiographical note records the exact time of birth as two-thirty in the afternoon.[2]

His parents, both aged twenty-four, had married earlier that year, on May 16.[3] We do not know whether he was a premature baby or whether he was conceived out of wedlock. But as a child, he was small and weak and frequently suffered from bad health, which argues for the former. Kepler himself understandably took the view that he had been conceived on May 17, the day after the wedding.[4] With remarkable, if somewhat implausible, accuracy he even gave the moment of conception as 4:37 in the morning.

At that time, Germany consisted of a patchwork quilt of largely independent states and cities within the Holy Roman Empire. In a prescientific and pre-Enlightenment age, religion was the dominant influence. Every German-speaking state was deemed to be either Catholic or Lutheran, depending on the religion of its ruler.

In 1555, rather than continuing to tear each other apart, the Catholics and Lutherans had reached a compromise known as the Peace of Augsburg. Under this complex and ambiguous agreement, every ruler of every state effectively had a free choice of which of these two religions to follow, with the proviso that his subjects could be required either to follow him in his choice or go into exile. This later became known as the principle of *cuius regio eius religio*[5]—whoever rules also chooses the religion. It was this agreement that kept religious peace in the German states for over sixty years, until the Thirty Years' War, which began in 1618 and which was to blight Kepler's final years.

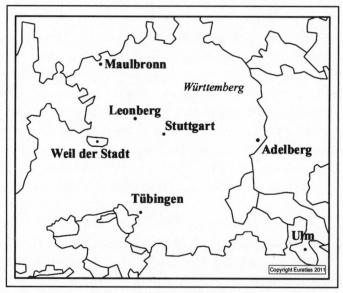

Figure 1.1. Württemberg in 1600 (Euratlas maps).

Weil der Stadt (one of the smallest of the eighty or so largely self-governing Free Imperial Cities) was in an anomalous position, as it was for the most part a Catholic enclave that was surrounded on all sides by the otherwise solidly Lutheran Duchy of Württemberg. (The records for 1590 show that there were 220 adult Catholic males and only 110 adult Lutheran males in the town.[6]) However, Lutherans were tolerated in Weil at that time, and Kepler was raised in the family tradition as a devout Lutheran.

Kepler's early years were characterized by a dysfunctional family, poor health, and the mutual hostility (real or imagined) between him and other schoolchildren. His lifelong search for harmony in the Universe was arguably, at least in part, a reaction against the total lack of harmony of his childhood years.

His father, Heinrich, was the chief villain in the family. Heinrich was the fourth child of Sebald Kepler and the first to survive infancy.

Kepler described him as "vicious, stern of character and headstrong" and firmly placed the blame on him for his parents' unhappy marriage, saying that "my father treated my mother very harshly." He did, however, explain Heinrich's violence and obstinacy as a consequence of the astrological influence of Saturn and Mars, and his malice as augmented by Venus and Mercury.

Kepler clearly had a little more respect for his paternal grandfather, Sebald, who had risen to become mayor of Weil, had achieved a great deal for the town, and was well liked by the town's nobility, one of whom left him a lot of money in his will. He had a "ruddy and fleshy face, and his beard gave him an appearance of authority." Kepler thought he was "eloquent for an uneducated man." But Sebald also had his downsides. He was "famous for his arrogance and pomposity in clothing" and was "irascible and stubborn, and his looks revealed he had been lustful." All this, Kepler explained, was due to the influence of Jupiter on him. As for Kepler's grandmother Katharina, she was "restless, clever, a liar, zealous about religious matters, of a fiery nature, vivacious, envious, inflamed with hatred, impetuous, and never forgot offences." He even suspected that—probably in order to get married—she had lied about her age, which he believed was really four years greater than she claimed.

His mother, also named Katharina, was the daughter of Melchior Guldenmann, an innkeeper from the nearby village of Eltingen and a magistrate in his community. She, too, came in for criticism; she was "small, thin, dark-complexioned, garrulous, quarrelsome and generally unpleasant." But "her stubbornness did not often overcome the inhuman behavior of her husband and her father-in-law."

Johannes was followed by two brothers and a sister: Heinrich, Margarete, and Christoph—three more children did not live to adulthood.[7] (Curiously, he makes little mention of his siblings in his main autobiographical notes, in spite of several references to his other relatives. Nor, in his nearly four hundred surviving letters or the nearly seven hundred surviving letters to him, is there any correspondence with his siblings. The relationship cannot have been particularly close.)

For most of the first four years of Kepler's life, his family lived with Heinrich's parents and some of Heinrich's brothers and sisters in a narrow-fronted house in one corner of the town's main square. (The house, together with much of Weil and most of the town's records, was destroyed in a fire in 1648, in the final stages of the Thirty Years' War. On its site now stands the Kepler Museum, built to the same plan as the original building and only a stone's throw from the Kepler monument that today dominates the square.[8])

Image 1.1. The Kepler Museum in Weil.

Image 1.2. The Kepler monument in Weil.

Kepler describes Heinrich as a wanderer who enjoyed fighting and vainly seeking honors. He was a mercenary soldier by profession and left home on a number of occasions to fight in wars elsewhere in Europe. In 1574, he went to fight in a war in the Netherlands. (And in spite of his Lutheran beliefs, he fought on the side of the Catholics against the Lutherans.) The following year, Katharina joined him, leaving Kepler to be looked after by his grandparents.

During that year, young Johannes fell seriously ill with smallpox. Fortunately for posterity, the disease at that time was not as virulent as it was later to become, and he survived. But it left him with badly crippled hands. And although he probably didn't realize it, it was almost certainly the smallpox that also left him with poor eyesight for the rest of his life; he suffered both from short sight and a defect that caused him to see multiple images, a cruel fate for anybody, not least an astronomer.[9]

Late in 1575, not long after returning from the war in the Netherlands, Heinrich gave up his citizenship of Weil, and the family moved to Leonberg, about 15 kilometers (about 9 miles) from Weil, in the safely Lutheran state of Württemberg. Their new home was close to Kepler's mother's original home in Eltingen and well away from his paternal grandparents. (The house can still be found at 11 Marktplatz in the old town's main square.) Heinrich then abandoned his family again to return to the war. Here he narrowly avoided being hanged, and at one stage his face was badly scarred by an exploding barrel of gunpowder. He finally returned to Leonberg, sold the house, and rented an inn in Ellmendingen, over 40 kilometers (25 miles) away. The family spent some two years there—interrupting Kepler's education with a period of agricultural labor—before eventually returning to Leonberg.

There were just a few good memories from this period. In 1577, when Kepler was five, his mother took him out to a high place one night to see the bright comet of that year. (This was the same comet that was also being observed in faraway Denmark by the thirty-year-old Tycho Brahe, who was concluding that—contrary to Aristotelian doctrine—it must be passing right through the crystal spheres that were meant to move the planets. Tycho correctly inferred that these spheres could therefore not be physically real.[10])

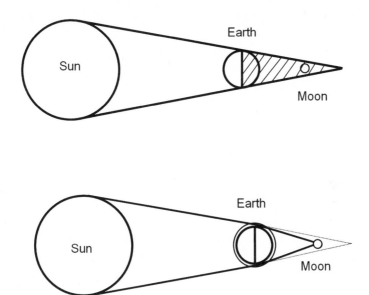

Figure 1.2. If the Earth had no atmosphere, then the moon would not be visible when it passed through the Earth's shadow (upper diagram). But any light not scattered by the atmosphere is instead refracted (lower diagram) and reaches the Moon's surface. This affects red light more than other frequencies, so the Moon appears reddish during an eclipse.

Then, one bitterly cold January night in 1580, his father called him outdoors to look at an eclipse of the Moon. During a lunar eclipse, the shadow of the Earth covers the surface of the Moon. But the Moon does not disappear completely. Some light from the Sun still reaches it because it is refracted through the Earth's atmosphere. Two effects are at work. First, light at the blue end of the spectrum is scattered by the Earth's atmosphere (which is also why the daytime sky looks blue). Second, the remaining light—at the red end of the spectrum—is not scattered but is refracted (bent) by the atmosphere, so it reaches the Moon's surface, which then appears reddish (fig. 1.2). This reddish Moon was what the young Kepler remembered. Many years later, he was to give the first largely correct explanation for the phenomenon.[11]

It was these two events (carefully recorded in his autobiographical notes) that sowed the seeds of a later interest in astronomy in Kepler, eventually leading him to become one of the two foremost astronomers of his day and one of the most respected astronomers of all time. For this, if nothing else, he did have something for which to thank his parents.

SCHOOL

In addition to the matter of religious doctrine, one of the complaints Lutherans had about Catholics was that a high proportion of their clergy were either poorly educated or not educated at all. The Lutherans were not going to fall into that trap. In Württemberg and elsewhere, Latin schools were set up to ensure that future Lutheran clergy and other boys received the best possible education.

Kepler's own abilities were soon recognized. So after a short time at a local German school, in 1577 he was sent to one such Latin school close to his home in Leonberg.[12] (The old school building still exists, but it is now a museum.) He took until 1583 to complete the course, which was interrupted by the family's move to Ellmendingen.

In 1584, he passed the *Landexamen* (state examination),[13] and in October of that year, at the age of twelve, he moved on from the Latin school to be a boarder at the seminary at Adelberg, for which—as the son of parents of modest means—he was given a grant from the Church. Here he continued his education and began training to become a Lutheran clergyman. This would have been completely in line with his own wishes, as he remained devoutly religious throughout his life. In this, he was very much a product of his environment; given his background of an unhappy and unstable family life, God and the Lutheran Church must have seemed an island of hope and meaning in an otherwise bleak existence.

The health problems that were to plague him for the rest of his life were very much in evidence at Adelberg. "During these two years, I suffered constantly from skin diseases, huge sores and badly healed wounds in my feet. In the middle finger of my right hand I had a worm, in my left hand a massive sore. In January and February 1586 I was constantly lying down and I was almost exhausted by treatments." He also had to put up with the hostility of other boys at the school: "Due to my illness, I had a bad reputation among the people of my age, who hated me." But he did well academically: "June 1586: It was exam time, and I earned praise."

Image 1.3. Kepler's Latin school in Leonberg.

After two years at Adelberg, in November 1586 he moved on to the higher seminary at Maulbronn, where he spent the next three years. The Maulbronn seminary had formerly been a (Catholic) Cistercian monastery, but the Lutheran Duke of Württemberg closed it down and, in 1556, turned it—along with another twelve former monasteries in his territory—into a Lutheran school. Of these thirteen former monasteries, Maulbronn was the jewel in the crown. The beautiful Gothic cloisters, the magnificent and spacious church, even the stately dining hall must have made a deep impression on the young Kepler. (Four hundred years later, Maulbronn's outstanding and well-preserved buildings have given it UNESCO World Cultural Heritage Site status.)

The training, though, was hard—the boys were woken between four and five in the morning, and lessons (in Latin) started an hour later. The focus was on theology and study of the Bible. But it also included the "seven liberal arts": grammar, rhetoric, logic, arithmetic, geometry, music, and astronomy, as well as lessons in Greek and Hebrew. Given the extreme intensity of the course, coupled with a meager diet, it is perhaps not surprising that Kepler's health problems continued. He reports that "in 1586, I had to endure such a hard time that I was almost consumed by my anxieties."[14] In a stressful environment, his relationships with his fellow students continued to be fraught: "I endured the hate of the people of my age, and with one of them I had a fight. . . . The majority of my schoolmates were enemies."

He still had a limited interest in astronomy. Hidden among his astrological musings, he records that in 1588 he saw another total eclipse of the Moon: "I was surprised, and I remembered the one I had seen in 1580."

SELF-PORTRAIT

In his midtwenties, Kepler composed a remarkably lengthy self-portrait (written in the third person). It gives a fascinating insight into how he saw himself during his years at school and beyond. He comes across as deeply introspective, neurotically obsessive, intensely pious, a hard worker, and—above all—decidedly precocious:

That man was born to this fate, to spend most of his time on difficult things that others hated. In his childhood, he attempted things
before he was old enough. He tried to write comedies, he picked
out the longest psalms, which he committed to memory.... He took
delight in puzzles, he sought the wittiest jokes, ... In writing about
problems, he loved to do so in paradoxes: the Gallic language was
to be learned before Greek, the study of literature was a sign of the
decline of Germany.[15]

His interest in mathematics and physics was already present at
school:

He loved mathematics above all other studies. In philosophy, he read
the original text of Aristotle for himself. He did research in physics.
... In mathematics he did much exploration in areas he thought had
not been explored before, and which afterwards he found out had
already been discovered. He devised a celestial clock.[16]

He also showed a keen interest in matters of religion. In an age
when belief in the Christian God was both universal and unquestioning,
it was simply a matter of choosing which particular version of this God
was the correct one. Kepler opted overall for the Lutheran version
in which he had been raised, but he had doubts and problems with
aspects of Lutheranism even from the age of thirteen:

In theology, he initially began with predestination, and fell in with
the Lutheran opinion of judgment. And amazingly, at the age of 13
years he wrote to Tübingen, so that a dispute over predestination
should be sent to him. He was taunted in a dispute on this: Rogue,
do you have temptations about predestination? Afterwards, he abandoned the view expressed in this work of Luther and moved to the
more acceptable view of [Aegidius] Hunnius [a Lutheran theologian].
Immediately, however, he got to grips with other Calvinistic contro-
versies. Furthermore, moved by consideration of the divine mercy,
he held the view that not all heathen nations would be completely
damned.[17]

He had an overdeveloped conscience:

He was pained by the fact that his devotion to vice meant that he was denied the gift of prophecy. ... If he had sinned, he made himself perform a penalty. Repentance was the recital of a particular sermon.[18]

He was something of a control freak as well as being constantly, and perhaps unnecessarily, concerned about money (a concern that was to haunt him throughout his life):

In money matters, he was too controlling, too inflexible in arrangements, criticizing the least little thing.[19]

The self-portrait also contained a lengthy list of people who Kepler disliked or who (he believed) disliked him.

Finally, he described himself as though he were a dog:

That man has in all respects the nature of a dog. He is like a spoilt pet dog. His body is agile, lean and well-proportioned. His way of life is the same as a dog. He likes to gnaw bones. He enjoys bread crusts. He is greedy, he tears up whatever he sees. He drinks little. He is satisfied with the smallest amount of food. ... He greets visitors like a dog. Whenever someone snatches something very little from him, he grumbles and gets annoyed like a dog. He tenaciously pursues those who act badly. He gets angry and is scathing towards others. He hates many people a lot. He shies away from baths.[20]

TÜBINGEN

Tübingen University had been founded in 1477, by Count Eberhart of Tübingen. Philip Melanchthon had studied and gone on to teach there. Duke Ulrich of Tübingen turned it into a Lutheran institution in 1534. Kepler won a scholarship and took up his place in the autumn of 1589 for what should have been a period of study of five years that should have culminated in his being ordained as a Lutheran minister of religion. (The year 1589 was also when his mercenary father went off to fight in yet another war and was never again seen by his family.[21]) All tuition costs for theology students were paid by the Duke of Württemberg,[22] and Kepler's scholarship meant that he obtained an annual grant for any further expenses. In addition, old Melchior Guldenmann, his maternal grandfather, was clearly anxious that his grandson should be as well provided for as possible and made over the income from one of his fields to the young student.[23]

The first two years were taken up with a continuation of the seven liberal arts subjects studied at Maulbronn, together with Greek and Hebrew.[24] The students were still taught in Latin (in which they were by then fluent), and their studies included the ancient Greek philosophers (to whom Kepler was later to make numerous allusions).

After the cloistered existence at Maulbronn, life at Tübingen provided limited opportunities to meet members of the opposite sex. In 1591, at the age of nineteen, he reports that he "suffered the pangs of love." Of the identity of the young lady concerned we know nothing, and there is no evidence that his feelings were reciprocated. A year later, we learn that "I was offered the hand of a virgin in marriage." The offer was not taken up, perhaps because he was still pining for his earlier love.[25]

While a student, he performed a part in a play, taking the role of Mariamne, the wife of King Herod. He seems to have been chosen for this particular role because of his small size. He lost money gambling. And he still had problems with other students: "Ortholphus hated me, as much as I hated Koellinus." His concerns for his health continued. "1590: I began to be seriously affected by headache and the disorder of my genitals. . . . The scabies weakened me. 1591: The cold brought out the scabies. . . . As I came out of church, I broke out in an intense sweat."

Academically, he made an excellent impression. He reports that at the end of his second year "I was second out of fourteen." He duly obtained his master's degree in August 1591 and was able to move on to the three-year course in theology. (Other students went on to study law or medicine.) His grant for his second year had already— exceptionally—been increased from 6 to 20 gulden per annum.[26] At the beginning of Kepler's third year, the Tübingen senate granted him a further 20 gulden, announcing that "because the above-mentioned Kepler has such a superior and magnificent mind that something special may be expected of him, we wish—for our part—to continue to that Kepler his stipend, as he requests, and because of his special learning and ability."[27]

MICHAEL MAESTLIN

But this rosy view of Kepler would not last. For it was at Tübingen that he met Michael Maestlin, the professor of mathematics and astronomy, and one of the very few people in the whole of Europe who recognized (in private, if not in public) that the Copernican system was correct. Maestlin was a talented man, although he had been lucky to obtain the post. For refusing to sign up to the Formula of Concord, his predecessor, Philip Apian, had been dismissed, and in 1584 Maestlin had been appointed in his place. (The Formula of Concord, produced in 1577, was an attempt to paper over the doctrinal cracks that were already appearing within Lutheranism. It asserted, among other things, that certain beliefs of the newly emerging Calvinists were false and heretical.[28])

Kepler attended Maestlin's lectures in astronomy and learned about the Ptolemaic system, which Maestlin taught as correct. Any other view, it was generally agreed, was contrary to the teachings of the Bible. But Maestlin also frequently mentioned Copernicus in his lectures. Copernicus was a respected figure, because his mathematical methods were the basis for the *Prutenic Tables* (even though his claim of a moving Earth was not believed), so Maestlin's references are not surprising. However, it is very unlikely that Maestlin publicly offered up Copernicus's Sun-centered view as an alternative version of the truth. Given the theological objections of the Lutherans, it seems far more probable that he only suggested the latter to some of his pupils privately.[29]

Kepler seems to have had an instinctive grasp for physical reality. Thanks to Maestlin, he saw very quickly that Copernicanism offered—at least in principle—considerable simplifications over Ptolemy. He became an immediate and enthusiastic convert:

> Already in Tubingen when I followed attentively the instruction of the famous Michael Maestlin, I perceived how clumsy in many respects is the hitherto customary notion of the structure of the Universe.

Hence I was so very delighted by Copernicus, whom my teacher very often mentioned in his lectures, that I ... repeatedly advocated his views in the disputations of the candidates. I already set to work also to ascribe to the Earth (on physical or, if one prefers, metaphysical, grounds) the motion of the Sun, as Copernicus does on mathematical grounds. For this purpose I have by degrees—partly out of Maestlin's lectures, partly out of myself—collected all the mathematical advantages which Copernicus has over Ptolemy.[30]

What he didn't know, and later complained about, was that if only he had known about a book called *Narratio Prima* (*First Account*), by Joachim Rheticus (the disciple of Copernicus who finally persuaded him to go public on his theory), he could have saved himself the bother of working it all out for himself.[31]

It was while he was at Tübingen, in 1593, that he first started writing about creatures that lived on the Moon. ("I enjoyed writing a dissertation on this topic as a joke.") A much expanded version of this would ultimately be published posthumously as *Somnium* (*The Dream*). In its final form, it was the story of a journey to the Moon, and it was the first relatively modern science fiction story ever to be written.

However, one aim of the book in its early stages was to counteract one of the arguments against Copernicanism. At that time, it was seen as an objection to Copernicanism that (according to Copernicus) the Earth is moving at enormous speed around the Sun, yet we are not conscious of any motion. Kepler's thesis pointed out that somebody on the Moon would not be aware of the Moon's motion and would see even more complicated planetary motions than we do on the Earth.

But he still intended to pursue the path of becoming a Lutheran clergyman. Then, completely by chance, something happened that changed the whole direction of his life. Late in 1593, only a few months before Kepler was due to complete his course, Georg Stadius, a mathematics teacher in his midforties at an obscure Lutheran school for boys in Graz (in faraway Styria) died.[32] The school authorities wrote to several universities[33] seeking a successor. If Tübingen University was to accede to the request, and if it was to pick one of its students, then Kepler was certainly a possible candidate, because he had shown that he was a talented mathematician.

However, there was no need for the university to respond positively or to choose Kepler. So it seems almost certain that two other factors were involved. First of all, Kepler was known to be an enthusiastic and public supporter of Copernicus, and the idea that Copernicanism was literally true was still frowned upon in the Lutheran Church.

Second, Kepler was known to have doubts over certain Lutheran doctrines, preferring certain aspects of the new sect of Calvinism. Throughout his life, he was never somebody to keep his views to himself. Many years later, he wrote about himself that "whatever I profess outwardly, that I believe inwardly. Nothing is a worse cross for me than ... to be unable to utter my inmost sentiments."[34] He frankly admitted that, in his early years, he "passionately discussed religion in public."[35] He also had "pangs of conscience [surely noted by the authorities] about chiming in with the frequent condemnation of the Calvinists."[36] Albert Einstein once described Kepler as someone who was incapable of doing anything but stand up openly for his convictions.[37]

These two factors explain the otherwise inexplicable: why the authorities at Tübingen did not even let Kepler finish the last few months of his course, why they ordered him to accept the post in Graz, and why they did not seem concerned that such a talented and exceptional pupil would be going to a lowly teaching post, rather than to a higher-status clerical position. It is otherwise impossible to understand their decision. It seems clear that they had decided (probably more in sorrow than in anger) to take this opportunity to quietly get rid of a brilliant student who had become an embarrassment.[38]

It was the twin themes of Kepler's Copernican ideas and his Calvinist streak that were to cause clashes with the Lutheran Church throughout the rest of his life.

Kepler objected strongly to the decision, "protesting loudly that I would never willingly concede my intention to follow another kind of life which seemed more splendid,"[39] as he still wanted to become a Lutheran clergyman. But the reality was that the Tübingen authorities gave him no choice in the matter. "However, to tell the truth, I was driven to take on this task by the authority of my teachers."[40] He was fully aware that he was being offered a low-status job, far away from his native land—Graz is nearly 700 kilometers (over 400 miles) from

Tübingen and was close to the border between the Holy Roman Empire and the Ottoman Empire. "I was not frightened by the distance of the place, . . . but by the low opinion and contempt in which this kind of function is held."[41] However, some time previously, he had noticed the reluctance of other students to move beyond the boundaries of Württemberg and had resolved never to be like them if he found himself in the same position.[42]

So, with no realistic alternative before him, he finally accepted the post and set off for Graz in March 1594. He was confident that one day he would return.

It was just as well for his state of mind that he did not know that the religious authorities would never allow him to work in Württemberg.

CHAPTER 2

GRAZ (1594-1600)—
THE MYSTERY OF THE UNIVERSE

The Sun's height in the sky at midday varies over the year. It is at its maximum height on the summer solstice (June 21, in the Northern Hemisphere), and at its minimum height on the winter solstice (December 21). A calendar based on the Sun needs to ensure that there is no gradual drift away from these dates. But there has always been a fundamental problem—there are not an exact number of days in the year. In fact, there are about 365¼, so unless a regular adjustment is made to the 365-day year, the date of the summer solstice will gradually change. If no regular adjustments were made, then after about 750 years, the summer solstice would fall on December 21.

This was recognized by the advisers to Julius Caesar, who introduced a year that was 365 days long for three years but 366 days long for the fourth—our leap year. Even this was not enough to prevent some drift in the calendar (because the precise length of the year is 365 days, 5 hours, 48 minutes, and 45 seconds, and not 365¼ days). So by the time Kepler was a boy, the summer solstice fell on June 11, rather than June 21. Pope Gregory XIII resolved to correct this. In 1582, he ruled that October 5 that year would become October 15. To avoid future calendar drift, he also ruled that the extra leap day would be omitted at the end of those centuries not divisible by 400. We still use this Gregorian calendar today.[1]

The Lutheran states were certainly not going to accept a papal decision that had undoubtedly been inspired by the devil! They stuck to the old Julian calendar. Confirmation that they were right came in the violent storms that raged over Germany during the ten days following October 5/15 that year—a clear sign of divine disapproval.[2]

So it was that Kepler set out on his journey from Tübingen on March

13, 1594,[3] but he did not reach Graz until April 21, 1594.[4] This was in part a reflection of the time taken for such a long journey in those days and probably also included time for a farewell visit to his family in Leonberg. However, he also lost ten days by the simple act of moving from a state ruled by a Protestant (Duke Friedrich I of Württemberg) to one ruled by a Catholic (Archduke Ferdinand of Styria). The ten-day difference was a source of embarrassment to him throughout his life, because he strongly supported the move to the new calendar even though he was a Lutheran. Many of his publications refer to both dates.

On his arrival in Graz, he almost immediately fell ill for a fortnight with what he described as Hungarian fever.[5] He finally began teaching late in May.[6] But teaching was not his strong point. If the Tübingen authorities really had thought he would be a good math teacher, they had shown decidedly poor judgment. In his first year, he had a few pupils at his lessons, and in the second year he had none at all. The school took a remarkably relaxed attitude to this, saying that "the study of mathematics is not everyone's meat."[7] Kepler excused his failure on the grounds that whenever he was giving a lesson, new thoughts and ideas were constantly popping into his head, and he was unable to stop himself from expressing them in whatever order they occurred. It is reminiscent of the old joke that time is the only thing that stops everything from happening all at once.[8] In Kepler's mind, time singularly failed to prevent his thoughts coming all at once. This, as he admitted, gave rise to his "ugly, confused and not very understandable language."[9]

He had other duties. As the district mathematician, he was required to produce an annual almanac giving an astrologically based prediction of events in the following year. Even though he had a belief in astrology, this was not a duty he enjoyed or had any faith in. He found the obligation to produce the annual almanac "burdensome,"[10] although he had to produce only six of them during his time in Graz. If nothing else, they were a useful way of earning a little extra money. However, in his very first forecast, for 1595 (which unfortunately we no longer have a copy of), he struck gold. He successfully predicted both the bitterly cold winter and invasions by the Turks.[11] This lucky guess greatly enhanced his prestige. (His contemporary Galileo was not so lucky in giving astrological forecasts. He once predicted that the

Grand Duke Ferdinand I of Tuscany would live a long and healthy life. The poor man died only a few months later.[12])

MYSTERIUM COSMOGRAPHICUM:
THE MYSTERY OF THE UNIVERSE

Since he had espoused the Copernican theory, three astronomical problems had been puzzling Kepler: why were there only six planets, why were they at the distances they were from the Sun, and why did they move more slowly in their orbits the farther they were from the Sun?[13] (The basis for the questions is set out in table 2.1—the planets Uranus and Neptune would not be discovered until 1781 and 1846, respectively).

Table 2.1. Relative distances, periods, and speeds of the planets that were known in Kepler's time

	Relative distance from Sun, taking Earth dist =1.0 (modern values)	Time to go once round Sun (modern values)	Speeds of planets (modern values)
Mercury	0.4	88 days	48 km/sec
Venus	0.7	225 days	35 km/sec
Earth	1.0	365 days	30 km/sec
Mars	1.5	687 days	24 km/sec
Jupiter	5.2	11.9 years	13 km/sec
Saturn	9.5	29.5 years	10 km/sec

These three questions were to form the underlying agenda for much of the rest of Kepler's life. There was no way he could have known that the first two questions were a complete waste of time, but that the third was one of the most important scientific questions ever to be asked. Yet it was his answers to the first two questions that were not only to come first but were also to provide the (totally mistaken) inspiration that was to motivate his later work.

What Kepler was to think of as his great insight came about initially as a result of the entirely accidental geometry of the solar system.

All the planets in the solar system orbit around the Sun in roughly the same plane. In other words, you can (with very little error) draw a diagram showing the Sun and the planetary orbits on a flat sheet of paper. There is no need to construct a three-dimensional model.

The Earth, of course, also orbits around the Sun in this same plane. But from our perspective we see this as the Sun moving against the background of the fixed stars and taking one year to return to its original position against this background. This path is called the ecliptic, and all the planets move very close to the ecliptic. Inevitably, therefore, there are times when they appear close to each other in the sky. When this happens, it is called a conjunction.

Jupiter orbits the Sun once every 11.9 years; Saturn takes 29.5 years. As a consequence, they have a close approach to each other—sometimes referred to as a Great Conjunction—roughly every 20 years.[14] (The next such conjunctions will be in December 2020 and November 2040.) Nowadays, we realize that these events are no more than attractive sights in the night sky, but in those days it was believed that conjunctions had some deep astrological significance. On July 19, 1595,[15] Kepler was explaining this phenomenon to his pupils (or at least to those who still bothered to come to his lectures) and explaining that—as seen from Earth—consecutive conjunctions occur (on average) a little less than 120° apart along the ecliptic, on the celestial sphere of fixed stars. He drew a rough version of the diagram in figure 2.1 to illustrate his point. (The large number of straight lines on the diagram are there simply to join up successive positions of the Great Conjunctions on the celestial sphere.)

It was then—in the middle of his lecture—that he had his revelation. What he (mistakenly) thought he had spotted was something completely different from what figure 2.1 shows. He had been pondering on the relative sizes of the planetary orbits for so long that he knew them all by heart.[16] In particular, he knew that the radius of the orbit of Jupiter was one-half of the radius of the orbit of Saturn (give or take a few percent—see table 2.1 above). The diagram in figure 2.1 (which had ended up looking rather like a succession of triangles drawn between two circles) sparked in his mind the thought that an equilateral triangle would therefore fit neatly and precisely between these two orbits (fig. 2.2).

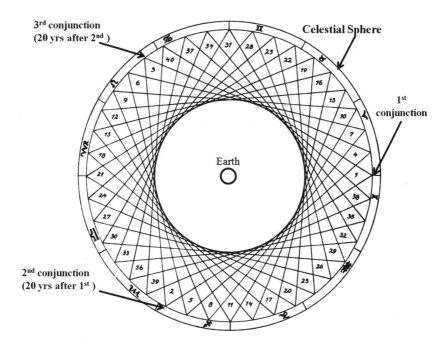

Figure 2.1. Annotated copy of Kepler's diagram of the positions of successive Great Conjunctions of Jupiter and Saturn in the night sky. The outer circle represents the celestial sphere. Successive conjunctions of Jupiter and Saturn occur roughly every 20 years and are roughly 120° apart in the sky.[17]

This was the moment he felt had come about by divine intervention.[18] This was his big idea. The idea was to be central in guiding him for the rest of his highly productive life and was one he never abandoned. It was a truly beautiful idea and the inspiration behind all his later work. Yet it was utterly and completely wrong.

The human brain can be thought of as a pattern-seeking device. Our success as a species is in part because of the ability of the brain to seek out patterns in nature. But, for sound evolutionary reasons, we often err on the side of seeing a pattern where none exists. (In evolutionary terms, for example, it is better to mistakenly see a tiger hiding in long grass, and to take evasive action, than to mistakenly fail to see the tiger and fail to take any action. In the first case—even though you have made an error in seeing a pattern that isn't there—you will live to pass on your genes. In the second, you failed to spot the pattern and you will not.)

Figure 2.2. If two concentric circles are drawn, in which one has twice the radius of the other, then an equilateral triangle will fit exactly between them.

Kepler's mistake—not for the last time in his life—was to see a pattern where none existed. Yet never before or since, in the whole history of science, has such a totally erroneous idea been the inspiration that has motivated an individual and led him on to such important breakthroughs.

The triangle was not literally there, of course, but it was present—Kepler thought—in the mind of God. Clearly, this was the basis on which God had designed the solar system nearly six thousand years earlier. The distances between the planets were not arbitrary. Each gap must contain a simple geometrical figure. Kepler was overwhelmed with excitement. Triangles would not fit between the other pairs of orbits, so he tried to find other simple two-dimensional shapes that would fit. He couldn't. However, by judicious choosing, he found he could instead achieve his purpose with three-dimensional shapes (the tetrahedron, the cube, the octahedron, the dodecahedron, and the icosahedrons—fig. 2.3).

There was more. It was believed that Pythagoras had been the first to discover these five "perfect solids" (so named because they each have a high degree of symmetry—all the sides of each solid are made of the same two-dimensional shapes), and Euclid had later proved that there could only be five perfect solids. So Kepler concluded that there were only six planets precisely because there were five perfect solids, which could fit neatly between the five pairs of orbits of the six planets.

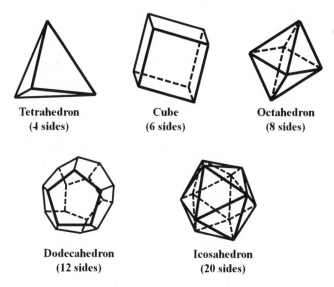

Tetrahedron   **Cube**   **Octahedron**
(4 sides)   **(6 sides)**   **(8 sides)**

Dodecahedron   **Icosahedron**
(12 sides)   **(20 sides)**

Figure 2.3. The five perfect solids.

> It will never be possible for me to describe with words the enjoyment which I have drawn from my discovery. Now I no longer bemoaned the lost time; I no longer became weary at work; I shunned no calculation, no matter how difficult. Days and nights I passed in calculating until I saw if the sentence formulated in words agreed with the orbits of Copernicus, or if the winds carried away my joy.[19]

The match turned out to be fairly close but certainly not exact, although Kepler thought it would be if he could only get hold of better data. This he knew was held by Tycho Brahe, the great observational astronomer.[20] The greatest discrepancy in one set of calculations was for Jupiter, which Kepler felt was at such a huge distance that nobody should really be in the least surprised.[21] This and other minor problems did not dampen Kepler's excitement. He had, at least to his own satisfaction, answered his first two questions.

But just how random was his choice for the order of perfect solids can be seen in table 2.2. If the solar system really had been constructed (by a mathematically inspired deity) in the manner Kepler thought, we might at least have expected that the solids would have been arranged in the order of the number of sides.

Table 2.2. Order of solids between planets.

Planet	Perfect solid between planets	Sides
Mercury		
	Octahedron	8
Venus		
	Icosahedron	20
Earth		
	Dodecahedron	12
Mars		
	Tetrahedron	4
Jupiter		
	Cube	6
Saturn		

Kepler did attempt to give a reason for the rather random order (by making use of the fact that the perfect solids can be divided into what are known as primary solids—the cube, tetrahedron, and dodecahedron—which are on one side of the Earth, and secondary solids—the icosahedron and octahedron—which are on the other side[22]), although it is deeply unconvincing. But the real reason was that this was the only way he could get them to fit (albeit not exactly).

The eager young Kepler rushed to write a book, *Mysterium Cosmographicum* (*The Mystery of the Universe*) setting out his discovery. One line of attack in the book is a forerunner of the style that he adopts, some twelve years later, in his *Astronomia Nova* (*The New Astronomy*). He tries to convince his readers that his final solution must be correct, if only because he has tried every other approach he can think of and has found them all faulty or inadequate. Like Sherlock Holmes, he believed that when you have eliminated the impossible, whatever remains—however improbable—must be the truth.

He records that, before his revelation and in his initial attempts to find whether there was a simple mathematical relationship between the distances of the planets from the Sun, he had tried inserting an extra planet between the orbits of Mars and Jupiter[23] (where there is a large gap—see table 2.1). Before abandoning the idea altogether, he had speculated that this planet might be too small to be seen. This was one of the many delightful occasions in his life when he later turned out to be right, albeit for the wrong reasons. Just over two hundred years later, in January 1801, the largest of the asteroids—Ceres—was discovered in the gap between Mars and Jupiter. Because it is so much bigger than the other asteroids in this gap, it has now been awarded the status (together with Pluto) of a "dwarf planet." But not all Kepler's guesses were correct He also speculated that there might be a small planet between Mercury and Venus. We now know that this is not the case.[24]

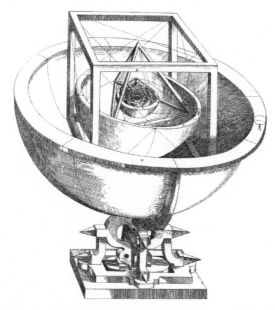

Figure 2.4. The illustration of Kepler's theory that appears in the *Mysterium Cosmographicum*.[25]

The introduction to the book shuns any suggestion of humility by proclaiming that the book reveals nothing less than the fundamental nature of the Universe and the reasons why God had structured it as it is.[26] There is a respectful reference to Pythagoras[27] who, two thousand years earlier, had attempted something similar, but who—Kepler informs us—just had not had Kepler's success.

Chapter 1 makes an overwhelming case for the truth of Copernicanism. One of Kepler's points is that nature prefers simplicity over complexity. If she can get away with a simple way of doing something, she will unhesitatingly choose that way, rather than something more complex.[28] His argument is a variant of Occam's razor—given two hypotheses with identical consequences, we should always prefer the simpler one to the more complicated one. The thrust of Kepler's claim is that the single, simple act of placing the Sun (rather than the Earth) at the center explains—rather than merely describes—everything that is observed, and does so in a simple manner. So, for example, Ptolemy in effect made use of five separate epicycles, one for each planet, to describe their retrograde motions, but Copernicus came up with a simpler theory by abolishing all these epicycles and introducing instead a moving Earth. This reduction from five motions to one motion not only explained the apparent retrograde motion of the planets, it also explained why the Sun was always exactly on the opposite side of the sky when an outer planet was in the middle of its retrograde motion (see figure I.6). And because the annual motion of the Earth caused it to move toward and away from each planet, this in turn explained small variations in latitude of those planets over the Earth's year. Ptolemy could also not explain why Mercury and Venus always remained so close to the Sun in the sky, and why (on average) they also took one year to orbit the Earth; in the Ptolemaic system, it was just yet another brute fact. But under Copernicus's system this was simply explained by saying that they were inner planets, orbiting more closely to the Sun than the Earth did. And last, the rotation of the Earth every 24 hours provided a much neater and more credible explanation of the apparent motion around the Earth of everything in the heavens than a fixed Earth. (In addition, Kepler throws in the fact that he has the support of at least one heavyweight—his old teacher Michael Maestlin.[29])

Most of the rest of the book contains highly dubious arguments that are principally concerned with a justification of his thesis on the five solids. He tries to give himself some wiggle room in his attempt to fit the perfect solids between the planetary orbits. He does this by widening the circles within which each planet orbits by making allowance for the fact that the planets vary their distance from the Sun. In what later turns out to be a critical move, he also takes the Sun, rather than the center of the Earth's orbit, as the center of the system. In addition, he widens the Earth's orbit by including the distance to the Moon—he frankly admits that he does this only to make the perfect solids fit better,[30] which it did for some planets but not others. And he cheats with Mercury, by allowing it to follow a different rule from all the other planets. His chapter 18 contains a candid admission that the fit is still not terribly good, but he blames this on poor observations rather than his own theory.

He introduces other claims as well. Having dismissed the Earth from the center of the Universe (about which he clearly felt a little concerned), he is anxious to make amends. So he explains that God had still (in what is, to put it mildly, a very obscure sense) placed the Earth in the center of the solar system, because the Sun, Mercury, and Venus were on one side, and Mars, Jupiter, and Saturn were on the other side. He introduces a series of obscure arguments to justify why his chosen order of regular solids was inevitable. He also puts forward astrological arguments about the nature of the planets and tries to justify the division of the ecliptic into twelve signs of the zodiac (which today we see as an entirely random choice) with appeals to numerological arguments derived from his perfect solids.

The final chapter (chapter 23) concludes that, based on the planetary positions he had calculated, the Universe must have begun 5,572 years previously[31]—to be precise, on April 27, 3977 BCE. (He acknowledges that others, who based their arguments more directly on biblical authority, had a good case that the correct figure was 5,557 years.) Unsurprisingly, neither of these figures would be accepted by a modern astronomer—all the latest observations unambiguously put the age of the Universe at 13.8 *billion* years.

But toward the end of the book, tucked away in chapter 20— apparently almost as an afterthought—another idea, something far more important and prescient, also creeps in. Finally, Kepler tries to answer the third of his three questions. He had been pondering the fact (mentioned in passing in Rheticus's book *Narratio Prima*[32]) that the planets move more slowly in their orbits the farther they are from the Sun (see table 2.1). What was the reason for this?

He answers his third question by putting forward the new and revolutionary concept that there was some sort of force coming from the Sun that was pulling the planets around in their orbits. He first points out that there could be only two possible reasons for the slower motion with greater distance from the Sun. Either each planet contained something (which at this stage he quaintly describes as a soul) that caused it to move, and the strength of each soul just happened to be weaker the farther it was from the Sun, or (far more likely) the cause of the planetary motions lay in the Sun and therefore became weaker with distance from the Sun.[33]

He speculates that the strength of this force reduced with distance in the same way as the intensity of light does.[34] His attempts at finding an exact mathematical relationship between distance from the Sun and time to orbit the Sun failed—this would not come for another twenty years—but the basic idea was there.

Kepler's chapter 22 expands on this theme. It was not simply that an outer planet moves more slowly than an inner planet. He had also noted that any particular planet moves more slowly when it is farther from the Sun and more quickly when it is closer to the Sun. He correctly recognizes that the explanation for this was simply that when the planet was farther away from the Sun, the Sun's force was weaker (so the planet moved more slowly), and when the planet was closer to the Sun, the Sun's force was stronger (so the planet moved more quickly).[35]

He even recognizes that this was the physical explanation behind the Ptolemaic concept of the equant. The equant was no more than a mathematical tool that tried to explain, in purely geometrical terms, why all the planets sometimes moved more quickly and sometimes more slowly. For the first time in human history, Kepler was suggesting that a physical force emanating from the Sun, and weakening with distance, provided a far better explanation.

Kepler had taken the first step toward answering his third question. This was the vital insight that would eventually lead to his three laws of planetary motion. His inspiration was his mistaken concept of the five perfect solids, but his revolutionary idea was that the Sun was exerting what we would now think of as a physical force on the planets, and that this force faded with distance. Something was needed to replace Aristotle's crystal spheres, which had been thought to move the planets in their orbits. It was Kepler who came up with the replacement. He was the first person ever to introduce physical forces into astronomy. He singlehandedly began the move of astronomy from mere geometry to physics. Even if he had achieved nothing else, this alone would justify his fame.

There were two key differences between Kepler and his contemporaries. First, he was one of the small minority who recognized the internal logic and the underlying simplicity of the Copernican system. But second, he was unique in asking questions about why the Sun-centered system operated as it did, and then in trying to find answers. His first two questions (on the number of planets and their relative distances from the Sun) turned out to be meaningless, even though his answers were what inspired him to greater heights. His third question (on why the planets moved more slowly as they got farther from the Sun) was critical to the development of modern science, and his answer was an important first step toward our modern understanding. Kepler rightly deserves his position as one of the giants of the scientific revolution.

PUBLICATION

In the twelve months following his discovery, Kepler wrote to Michael Maestlin no fewer than ten times about it. Early in 1596, he was given leave by the school to return home for a couple of months to visit his elderly and ailing grandfathers. This also gave him the opportunity to visit Maestlin and discuss the book with him. He also wasted much time in negotiations with the Duke of Württemberg, in Stuttgart, on the creation of an elaborate model of his idea (which would have been a huge version of the illustration shown in figure 2.4). This eventually came to nothing. But before the book could be published, he first had to get the permission of Tübingen University, which he obtained in June.[36] This was in no small part due to the efforts of Maestlin, who wrote to Matthias Hafenreffer, the head of the university, that

> the topic and the ideas [in the book] are so new that up to now they never entered anybody's mind. ... For whoever conceived the idea or made such a daring attempt as to demonstrate a priori the number, the order, the magnitude and the movements of the celestial spheres.[37]

Maestlin's only criticism of the book was that it did not give a sufficient explanation of the underlying Copernican system on which it was based.

Kepler finally returned to Graz in July 1596,[38] leaving poor old Maestlin with the burden of getting the work published.[39] Maestlin took the opportunity to improve the explanation of the Copernican system by adding, in an appendix, the *Narratio Prima* by Copernicus's disciple Rheticus. Back in Graz, the authorities—perhaps surprisingly—overlooked Kepler's extended period of absence.

There were still hurdles to be overcome before the book could be published. Kepler had originally intended to include in the book an explanation of why he thought there was no conflict between Copernican theory and the Bible. This, however, was refused by the authorities at Tübingen University, who did not want to stir up trouble on such a sensitive issue. The printers had been frightened of taking a loss, and those who opposed the Copernican theory, he tells us, did their best to interfere.[40]

The position of the authorities at Tübingen University toward Kepler's book seems at first sight inconsistent. On the one hand they sanctioned the publication of an avowedly Copernican book (apparently implying that they had no objection to the Copernican system). On the other hand, they were very clear in instructing Kepler that he was not to include any attempt to reconcile the (Earth-centered) teaching of the Bible with the Copernican view (showing that they believed that the two were incompatible with each other). The explanation of these apparently contradictory positions is simply that the Tübingen authorities took Andreas Osiander's instrumentalist view that the Copernican system was merely a calculating device, and was not to be considered as objectively true.

The attitude of the Tübingen authorities is exemplified in the view of Matthias Hafenreffer, as reported by Maestlin. While on the one hand regarding Kepler's book as "outstandingly imaginative and learned," he nevertheless thought it was "quite contrary to the truth of Holy Scripture."[41] Hafenreffer also wrote directly to Kepler urging him not to suggest that Copernicanism was true but just to treat it as a means of calculating planetary positions. Otherwise, he said, "many good people would take offence, and not unjustly."[42] Kepler, in his usual optimistic way, thought Hafenreffer was a secret supporter of Copernicanism, although Maestlin's comments make this deeply unlikely.

But *Mysterium Cosmographicum* was at last published early in 1597 and was registered in the Frankfurt catalog—in which, to Kepler's annoyance, his name was misspelled as Repleus.[43] It is noteworthy that it was almost the first book, fifty-four years after the publication of Copernicus's book, to come out publicly in favor of the Copernican Universe—or, more accurately, Kepler's version of it. (He was beaten to it by Giordano Bruno and Thomas Digges.) Kepler circulated the book widely, sending it to some of the most famous names of the day and gaining a reputation as a bright young man with interesting ideas, even though the book did not receive the rapturous reception he had perhaps hoped for. His statement, twenty-five years later, that the whole of *Mysterium Cosmographicum* was immediately recognized by the astronomical community as both important and true[44] again reveals his tendency to put an optimistic gloss on events—in fact, hardly anyone agreed with it.[45]

Three years after its publication, however, in 1600, Kepler was finally paid 200 thalers (more than one year's salary) for the book from the noblemen of Styria, to whom he had dedicated it.[46]

GALILEO

One copy found its way into the hands of a relatively unknown professor of mathematics at Padua University by the name of Galileo Galilei

(1564–1642).[47] The messenger who had brought him the book was about to return to Graz, so Galileo lost no time in dashing off a letter to Kepler enthusiastically endorsing Kepler's Copernican approach:

> So far I have read only the introduction, but have learned from it in some measure your intentions and congratulate myself on the good fortune of having found such a man as a companion in the exploration of truth. For it is deplorable that there are so few who seek the truth.... I would certainly dare to approach the public with my way of thinking if there were more people of your mind. As this is not the case, I shall refrain from doing so. The lack of time and the ardent wish to read your book make it necessary to close, assuring you of my sympathy.[48]

It is this letter that tells us that Galileo was, even as early as 1597, a closet supporter of Copernicus. His modern-day detractors have sometimes alleged that he took a delight in deliberately stirring up trouble, and there probably was a streak of this in his character. But the letter shows that, far from wanting to inflame matters, Galileo (like Maestlin) was determined to restrict his Copernican ideas to a very limited circle. It was only thirteen years later, after his telescopic discoveries had turned up solid observational evidence backing the Copernican view, that he would finally take the risk of going public.

Kepler was delighted by Galileo's response and wrote to Maestlin telling him so. He also replied to Galileo:

> I received your letter of 4th August on 1st September. It was a double pleasure to me. First, because I became friends with you, an Italian, and second because of the agreement in which we find ourselves concerning Copernican cosmography. ... I am sure, if your time has allowed it, you have meanwhile obtained a closer knowledge of my book. And so a great desire has taken hold of me, to learn your judgment. For this is my way, to urge all those to whom I have written to express their candid opinion. Believe me, the sharpest criticism of one single understanding man means much more to me than the thoughtless applause of the great masses.[49]

Galileo did not reply to this. Probably when he finally read Kepler's book, he concluded that its mysticism and numerology took the wrong approach. One of the tragedies of the era was the fact that, although Kepler and Galileo were fighting on the same side, they had different temperaments and different ways of tackling problems. Galileo was the pragmatist, with his feet firmly on the ground. Albert Einstein rightly described him as the father of modern science. But Kepler had a free-ranging imagination that sometimes led him to the truth and sometimes into the realms of fantasy. They never had the meeting of minds that one might have hoped for. And yet for science to advance, it probably needs both dreamers (Kepler) and skeptics (Galileo).

NICOLAUS REYMERS URSUS

Kepler also sent one copy of the book to Nicolaus Reymers Ursus (1551–1600), who was then the Imperial Mathematician to the eccentric Holy Roman Emperor Rudolph II in Prague. Ursus was a shady individual, who had once stayed as a guest of Tycho Brahe at his observatory of Uraniborg in Denmark. Tycho had been writing about his proposed alternative to the Copernican system (described in the introduction to this book), and Ursus may have found and read the manuscript without permission. Sometime later, he published details of his own alternative system, which bore a remarkable resemblance to Tycho's. Whatever the truth of the matter (because it is certainly possible that Ursus arrived at much the same ideas independently[50]), Tycho never forgave him. He took his revenge in 1596, by publishing accusations of plagiarism and theft against Ursus.

Kepler had no idea of the extreme level of enmity between the two men. In all innocence, he had written to Ursus late in 1595, setting out his ideas for *Mysterium Cosmographicum* and asking for Ursus's opinion. The letter, in typical Kepler style, heaped praise on Ursus, who (not one to miss a trick) chose to publish Kepler's letter in a preface to Ursus's own book, in order to show the high regard in which he was held. Ursus's book was also heavily critical of Tycho's attack on him. Tycho was sent a copy of the book, and Kepler's letter of praise did not help him—to put it mildly—in his relationship with Tycho.

TYCHO BRAHE

Tycho Brahe was a colorful character.[51] He had been born into an extremely wealthy Danish family and had been kidnapped by his (childless) uncle at the age of two. His uncle then raised him as his own, apparently with no objections from Tycho's parents. He became interested in astronomy at the age of thirteen when he saw a partial eclipse of the Sun. What fascinated him was not primarily the spectacle of the eclipse but the fact that it had been possible to predict it in advance. He eagerly took up the subject but quickly became dissatisfied with the lack of precision in forecasts of planetary positions. He resolved to devote his life to making more accurate observations.

Shortly after his twentieth birthday, he lost his nose in a duel. (The whole history of astronomy, and probably of all the physical sciences, would have been totally different if he had lost his eyes instead of his nose. It must have been a close thing.) Replacement noses at the time would normally have been made of wax. Tycho was more than rich enough to afford a replacement made of a gold/silver alloy. It has been widely believed that this was the case for at least one replacement— perhaps one reserved for special occasions. This would not just have been out of vanity (although this would undoubtedly have been a factor); gold and silver react less than other metals with human skin.

In 1572, he observed a bright new star (which we now know was a supernova) that remained visible for the next eighteen months. He demonstrated by its lack of movement relative to the fixed stars that it must lie beyond the sphere of the Moon. This was a blow to the followers of Aristotle, who held that everything beyond the Moon's orbit was perfect and unchanging.

With some financial help from the king of Denmark, Tycho set up the magnificent observatory of Uraniborg in 1576, on the Danish island of Hven. He then spent the next twenty years making by far the largest and the most accurate set of observations of stellar and planetary positions over time that had ever been made. His instruments were pre-telescopic (the telescope would not be invented until a few years after his death), but they measured stellar and planetary positions with an astonishing and unprecedented accuracy of two minutes of arc (two-sixtieths of a degree) or better.

In 1577, he observed a particularly bright comet and concluded that the comet's orbit must pass through the orbits of some of the planets. This demolished the Aristotelian view that the planets were moved in their orbits by gigantic transparent crystalline spheres. It also created a need to find some other mechanism for moving the planets—a point on which Kepler was to alight in suggesting that the mechanism lay in a force emanating from the Sun.

In 1588, he published a book setting out, among other things, his compromise system—halfway between that of Ptolemy and that of Copernicus—in which the Earth remained stationary at the center of the Universe. The Sun (and the Moon) orbited the Earth, but the planets orbited the Sun. He claimed in the book that he had first thought of this system in 1583. Ursus had visited Tycho at Uraniborg in 1584, in his capacity at the time as a servant to a Danish nobleman named Erik Lange, and went on either to plagiarize Tycho's work or to arrive at much the same ideas independently.

Tycho maintained a lavish lifestyle at Hven. As well as all his observatory staff, he kept a dwarf who was said to be able to foretell the future. He also kept a pet elk, which—alas—got drunk one day while in the care of a relative, fell down some stairs, broke its leg, and died. He was a loyal family man, although he never married the woman who bore his eight children—he was a nobleman and she was a commoner, so marriage was out of the question. He had a regular flow of important visitors to Hven, including—in 1590—King James VI of Scotland (later to become James I of England). But it was his treatment of the peasants on the island of Hven, together with the unwillingness of the new king Christian IV to continue paying for the high costs of the observatory, that were his undoing. Eventually—in 1597—he was forced to leave.[52]

He and his family spent the next two years moving steadily southward. His first stop was his home in Copenhagen. He then journeyed on via Rostock to Wandsbeck, near Hamburg, arriving there in October and staying for almost a year. While in Wandsbeck, he started to make overtures to Rudolph II, in the hope of finding employment with him. He spent the winter months of 1598-99 in Wittenberg, where he lodged in the same house as Johannes Jessenius, a professor of medicine, with whom he became good friends. In June 1599, he finally moved to Prague to take up the post of Imperial Mathematician to Rudolph.

In December 1597 Kepler sent a copy of his book to Tycho, but it was some months before Tycho received it. He replied fairly promptly[53] in April 1598, but there was a delay of nearly a year before Kepler received the response. Tycho's reply was courteous and encouraging, notwithstanding his undoubted fury at Kepler's praise of Ursus.

> I like [your book] in no ordinary degree ... your genius shines forth in unambiguous terms ... ingenious and succinct ... to make use of planet distances and the symmetries of the regular bodies.[54]

Much of the rest of the letter set out observational details that Tycho felt were in conflict with Copernicus's and Kepler's theories. But crucially, and in spite of his criticisms, Tycho hinted that he could help Kepler in his work if he were to pay a visit. However, in a lengthy postscript, Tycho railed against Ursus and the fact that Ursus had made use of Kepler's letter:

> The same messenger who brought your letter to me ... brought to me there at the same time ... the notorious and abominable writing of a certain Ursus, more a wild beast than a man. When I read through its impudent calumnies and the unspeakable lies and insults in which it everywhere abounds altogether shamelessly and beyond measure, I found there a certain letter of yours with which he attempts to adorn himself and hide his shame.[55]

The postscript continued with a lengthy condemnation of Ursus. At the same time, Tycho also wrote to Maestlin in an even more critical tone, complaining both about Kepler's a priori approach to astronomy and his praise of Ursus. Maestlin received this letter in August and duly alerted his old pupil to his faux pas:

> I understand [from Tycho] that Ursus has published a certain book in which the most offensive jibes are directed at Tycho, to which he prefaced a letter of yours in which you honor him in the most glowing terms. I indeed have not seen the book, nor can I believe that

you wrote such a letter. For you know my opinion of that man. The things he published in his book are not his own, nor did he understand them properly, so that what is good in the book he expresses in the wrong words. He derives many things from Tycho and retails them as his own.[56]

As late as February 1599, Kepler finally received both Tycho's letter to him and a copy of Tycho's letter to Maestlin. He immediately sent a lengthy apology to Tycho. (All Kepler's letters were lengthy—he didn't do brevity.) The apology disowned both Ursus and his own earlier letter:

> There is now, surely, no dispute about Ursus' nature; undoubtedly, at the time I wrote the letter I was more foolish in my judgment, and I would not still pay the same tribute to him.[57]

Having received a hint of the possibility of future employment with Tycho, he was determined not to let it slip away because of his earlier blunder.

MARRIAGE

It was also in 1597, shortly after the publication of his book, that Kepler married Barbara Müller, the daughter of a wealthy property owner. We first hear of her in Kepler's autobiographical notes, when he reports that he had been offered a wife shortly after arriving in Graz.[58] After lengthy negotiations and Kepler's long absence in Württemberg (which nearly wrecked matters altogether), the issue was finally settled with their engagement in February 1597. They were married in April "under a calamitous sky"[59]—Kepler's first suggestion that the marriage was not to be entirely happy. Although Barbara was only twenty-two, she had already been married and widowed twice, and she brought a young daughter, Regina, to the marriage.[60] The couple moved into their new home at 6 Stempfergasse (now in a fashionable shopping area in the center of the old town of Graz). Kepler's relationship with his new father-in-law was a stormy one. Jobst Müller treated him with contempt and derision, or so it seemed to Kepler's imagination at the time.[61]

Less than a year later, in February 1598, their first son, Heinrich, was born. Kepler cast a horoscope promising a wonderful future for the child—he would have a noble character and would be nimble, mathematically inclined, and diligent. But when he was only two months old, Heinrich died.[62]

FAREWELL TO GRAZ

However, religious strife (the bane of much of Kepler's life) was about to break out. Twenty years before Kepler had arrived in Graz, almost all the inhabitants of the city had been Lutherans. Its ruler, Archduke Charles II, was a Catholic. In 1573, he founded a Jesuit college—later upgraded to a university—in the city and did his best to weaken the hold of the Lutherans. He died in 1590 but made his son and successor Archduke Ferdinand (then only twelve) swear to maintain Catholicism in the city.[63]

Ferdinand's whole life was a classic example of the truth of the Jesuit boast: "Give me a child for seven years, and I will give you the man." In Ferdinand's case, the indoctrination process took only five years. In 1590, he was sent to the Jesuit University of Ingolstadt in neighboring Bavaria and the Jesuits' main base in southern Germany. Here he received his education, returning to Graz in 1595 as a devout Catholic to begin both his reign and his lifelong persecution of Protestants.

In March 1598, there was a partial solar eclipse in the constellation of Pisces. Kepler describes this as a highly significant event for Ferdinand and something that was a presage of the troubles that were about to descend on Graz.[64] Sure enough, in September of that year, Ferdinand issued a decree that all Lutheran preachers and teachers were to leave Graz. When it was pointed out to him that such expulsions risked causing economic disruption, he replied that he would rather rule a country ruined than a country damned.[65]

Kepler was among those thrown out of Graz (with no choice but to leave his wife and stepdaughter behind), but he was alone in being allowed back, only a month later. It is not clear why he was given this exceptional treatment. Perhaps it was because he also had an official function as district mathematician, but more likely it was because he had friends in high places. It may be that these friends hoped he would convert to Catholicism. If so, they were to be disappointed. It has been suggested that one Catholic friend in particular, Herwart von Hohenburg, an important Bavarian diplomat with a deep interest in astronomy (and with whom Kepler frequently corresponded on astronomical matters) was responsible for Kepler being allowed back into Graz. It might conceivably have been Ferdinand himself who was responsible, as he had some interest in scientific matters. We shall never know.[66]

However, it was increasingly clear that no Lutheran, however prominent, would be able to remain living in Graz for much longer. Persecution of the remaining Lutherans was growing steadily more intense. Kepler's second child, Susanna, was born in June 1599 and died thirty-five days later.[67] He had to pay a fine of 10 thalers (reduced to 5 on appeal) to bury her in a Lutheran cemetery. Books deemed to be heretical were sought out and destroyed. The Lutheran churches were closed down. And the school where he taught had also been closed,[68] although his employers continued to pay his salary.[69] He needed to find an escape route. He tried, via Maestlin, to get a job at his old university in Tübingen. Although he was still a devout Lutheran, he no longer had any desire to be a clergyman. Instead he was looking for a "professorship of philosophy or even medicine."[70] But Maestlin was unable to help.

Tycho, meanwhile, had been in Prague since June 1599 in his capacity as the new Imperial Mathematician to Rudolph II. Prague was not such an impossibly distant place as Hven or Wandsbeck, so with no other realistic choice open to him, Kepler finally decided that he should pay a visit to Tycho. He was still keen to get hold of Tycho's observations, to obtain better figures for the sizes of the planetary orbits so as to confirm the ideas in *Mysterium Cosmographicum*. Early in 1599, he had written to Maestlin:

> My opinion of Tycho is this: he has abundant wealth. Only, like most rich men, he does not know how to make proper use of his riches. Therefore, one must take pains to wring his treasure from him, to get from him by begging the decision to publish all his observations without reservation.[71]

He traveled to Prague (setting out on January 1, 1600) courtesy of Friedrich Hoffmann, a friend and Styrian nobleman, who was also going to Prague and who offered him a lift.[72] Hoffmann also provided Kepler with temporary accommodation in Prague. He left Graz just before a second letter arrived from Tycho, asking Kepler to come and join him. The letter yet again took the opportunity to launch another attack on Ursus and also urged Kepler to drop his a priori speculations (in deciding on a theory before having looked at all the observational evidence), and instead urged him to proceed on an a posteriori basis (evidence first, theory afterward).[73] Kepler seems to have taken this advice to heart in his later work, although his speculative tendencies never left him.

JOHANNES KEPLER
1571 — 1630

LEHRTE HIER AN DER EINSTIGEN PROTESTANTISCHEN
STIFTSSCHULE 1594 – 1599 ALS PROFESSOR FÜR MATHEMATIK.
IN GRAZ SCHRIEB ER SEIN ERSTES ASTRONOMISCHES WERK
"DAS GEHEIMNIS DES WELTENBAUES",
DAS IHN IM GANZEN ABENDLAND BERÜHMT MACHTE.
1600 MUSSTE ER IM ZUGE DER GEGENREFORMATION GRAZ
VERLASSEN UND WURDE AM HOFE KAISER RUDOLF II. IN
PRAG MITARBEITER UND NACHFOLGER TYCHO DE BRAHES.
DIE EVANGELISCHE STIFTSSCHULE WURDE GESCHLOSSEN
UND IN EIN KLOSTER DER KLARISSINNEN UMGEWANDELT.

Image 2.1. The plaque commemorating Kepler's time as teacher in Graz.[74]

Four hundred years after Kepler and the other Lutherans were first thrown out of Graz, the city was designated a UNESCO World Cultural Heritage Site because of its remarkably well-preserved and beautiful old town. But the school building where Kepler taught has not survived. After it was closed, Ferdinand had it converted into a convent. On the site now stands a large department store. Only a grubby plaque on one of its walls reminds the visitor that this is where Kepler had the inspiration that was to shape the rest of his life.

CHAPTER 3

KEPLER AND TYCHO

B enatky Castle lies about 40 kilometers (about 25 miles) to the northeast of Prague.[1] It had been placed at Tycho's disposal by the Emperor Rudolph II to enable him to continue his astronomical observations. Situated on the top of a hill, it has a commanding view of the surrounding countryside. It was here that one of the most important encounters ever in the history of astronomy took place.

Image 3.1. The courtyard at Benatky Castle.

Kepler arrived in Prague late in January 1600, a little over six months after Tycho. When Tycho heard the news of Kepler's arrival, he wrote a letter of welcome: "You will come not so much as guest, but as a very welcome friend and highly desirable participant and companion in our observations of the heavens."[2] The two met at Benatky early in February.[3] It was a meeting of opposites who needed each other. They had in common only their genius and their deep interest in astronomy. Otherwise, they were worlds apart. Tycho was a rich and aging nobleman, almost twice Kepler's age, with a domineering personality and a deep sense of his own superiority. The younger Kepler had come from a much humbler and less secure background. Tycho was primarily an observer, but Kepler was more of a theoretician. Tycho wanted to use Kepler's mathematical skills to help to demonstrate the truth of his Tychonic model of the Universe and probably also to help him exact revenge on Nicolaus Reymers Ursus. Kepler simply wanted to get hold of Tycho's observations in order to verify his own version of the Copernican theory. Tycho initially saw Kepler as just one member of his group of observers, but Kepler was not really a team player. The possibility of conflict was there from the start.

Kepler soon became deeply unhappy with conditions at Benatky. His main motive for coming to Tycho had been to get hold of the latter's observations in order to confirm the theory set out in *Mysterium Cosmographicum*. But he was not given access to all Tycho's observations—only those Tycho chose to toss out to him over the noisy and interminable communal meals he was required to attend.[4] Tycho was not going to allow access to all his valuable observations, especially to somebody so new to his team, and someone who—to Tycho's suspicious nature—might possibly still have some link to the hated Ursus.

Nor was Kepler given a formal post. This, plus the strain of working in a chaotic environment with so many others, rather than being his own master, eventually became too much. He felt he needed clarification of his position. To make his period of collaboration with Tycho worthwhile, he needed to stay for a year or two. But how were the details of this to be arranged? He wrote a long and detailed letter to Tycho, setting out eleven separate problems.[5] How could he obtain the consent of his employers in Styria to such a long absence? Should Kepler seek it? Should Tycho? Should the emperor? If permission were given, his wife would have to join him. But where should they stay? And Tycho would have to agree to supply firewood to the household, as well as meat, fish, beer, wine, and bread. Tycho would also have to allow Kepler time off during the day if he had been up for most of the night observing. And he would have to pay Kepler fifty florins a quarter for his work. And so on.

A formal meeting between the two was held in April to try to resolve matters. It was also attended by Johannes Jessenius, who had met Tycho in Wittenberg. Jessenius had moved to Prague and was present to act as a neutral observer. He was later to become a good friend of Kepler.[6] The meeting did not go particularly well, in spite of Tycho's conciliatory attempts, and Kepler left for Prague in a sulk. Here he wrote Tycho a bitter letter, which simply had the effect of alienating him.

But Kepler soon realized what a foolish mistake he had made, wrote to Tycho again (in a tone that can best be described as groveling) begging his forgiveness, and was received back into the fold. Arthur Koestler, in *The Watershed: A Biography of Johannes Kepler*, paints a delightful picture of the reconciliation, with a jovial and avuncular Tycho hugging a chastened and diminutive Kepler.[7] Tycho agreed to do his best to persuade Emperor Rudolph to agree that Kepler should work for him for the next two years, provided that Kepler's employers in Graz were in turn willing to continue to pay Kepler's salary.[8]

RETURN TO GRAZ

In June, Kepler returned to Graz (via Vienna[9]) in the company of a relative of Tycho, Friedrich Rosencrantz[10] (probably the same Rosencrantz who was later to be immortalized by William Shakespeare and Tom Stoppard). Having reached an agreement of sorts with Tycho, he now intended to collect his wife and stepdaughter "and my bookcases,"[11] and settle his affairs there.

But back in Graz, things did not go smoothly. His employers told him that he should abandon astronomy and instead go to Italy to take up the study of medicine. Far better, they argued, to do something useful than something merely interesting. He certainly could not expect them to continue paying his salary unless he did this. This initially seemed to ruin any chance of returning to Prague, as Kepler's position there was conditional on his continuing to receive his Graz salary. He toyed with the idea of working for the Archduke Ferdinand as a mathematician, but this came to nothing.

Some of his correspondence when back in Graz revealed his feelings about his relationship with Tycho:

> I would have concluded my research on the harmonies of the world if Tycho's astronomy had not fascinated me so much that I almost went out of my mind. . . . One of the most important reasons for my visit to Tycho was the desire, as you know, to learn from him more correct figures for the eccentricities in order to [compare them with] my *Mysterium*. . . . But Tycho did not give me the chance to share his practical knowledge except in conversation during meals, today something about the apogee, tomorrow something about the knots of another planet.[12]
>
> Tycho possesses the best observations and consequently, as it were, the material for the erection of a new structure; he also has workers and everything else which one might desire. He lacks only the architect who uses all this according to a plan. For even though he also possesses a rather happy talent and true architectural ability, still he was hindered by the diversity of the phenomena as well as by the fact that the truth lies hidden exceedingly deep within them. Now old age steals upon him, weakening his intellect and other faculties or, after a few years, will so weaken them that it will be difficult for him to accomplish everything alone.[13]

ECLIPSE

While in Graz, he was able to carry out observations in the town's main square of the partial solar eclipse of July 10, 1600. To do this, he had constructed the ingenious wooden device shown in figure 3.1.

A pinhole at the top of the diagonal staff created a projection of the Sun at the bottom, and the image could be clearly seen if the whole device was covered in "so many layers of black cloth that no light can break in."[14]

The angle of the staff could be varied both horizontally and vertically to maintain the image as the Sun moved across the sky.

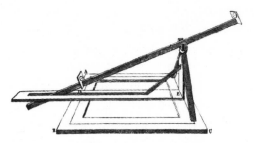

Figure 3.1. Kepler's eclipse-viewing device.[15]

FINAL EXPULSION

Events then forced Kepler to make a decision. Ferdinand had finally had enough of the remaining Lutherans. In a symbolic gesture, some ten thousand heretical Protestant books were publicly burned.[16] In late July 1600, a decree was issued that all the inhabitants of Graz were to attend a hearing at which their religious beliefs were to be examined. Anyone at this hearing who was not a Catholic, and who did not agree to become a Catholic, was banished. Kepler was among those who refused to convert, in spite of rumors to the contrary, and was given just over six weeks to leave.

He immediately wrote to Tycho, who promptly responded that Rudolph had given permission for Kepler's appointment as one of his assistants and urged Kepler to come and join him.[17] But Kepler was clearly still very much in two minds about the move to Prague and still longed to return to his homeland. So he sent a final plea for a position at Tübingen to Maestlin:

> I would go to you with my family by boat on the Danube, if God lets me survive. I would start a medical career if you would perhaps give me an assistant professorship. For indeed, I have become quite poor, after having hoped to become rich one day. I have taken a wife of good fortune; her whole family is in the same position. Yet all her possessions are invested in property, and since these have come down in price now, they are hardly marketable. Everybody is lying in wait to get them for nothing.[18]

In September, he then set off for Prague with his wife, his stepdaughter, and two wagonloads of all their possessions. The journey must have been a wretched one. He was struck down by quartan fever, a relatively benign form of malaria that results in fever every third day (or every fourth day counting inclusively, which explains the name). The fever continued intermittently for most of the next year.[19] On his arrival in Prague, his friend Baron Hoffmann again gave him accommodation. During his absence, Rudolph had decided to relocate Tycho from Benatky to Prague.

His receipt of Maestlin's eventual reply, in December, informing him of the lack of any job prospects at Tübingen University pushed him into a state of complete despair:

> As I succumbed to the quartan fever, I cannot tell you what a paroxysm of melancholy your letter caused me, by destroying all my hope of being appointed at your university. Here in Prague everything is uncertain, and uncertain with regard to my life too, but now I must stay here until I either recover or die. Everything is four times more expensive here than elsewhere. I especially pity my wife, who is with me. . . . I entreat you, venerated teacher, to insist on my being appointed [at Tübingen] as soon as there is any vacancy. I certainly would conduct myself as befits a grateful pupil. Believe me, if I came to Tübingen many noble men of Styria who would [otherwise] go to other places would study there.[20]

Kepler wrote again to Maestlin early in 1601 but didn't receive a reply to either letter. Perhaps Maestlin was simply growing tired of his former pupil's constant demands, but he didn't reply to any of Kepler's letters for another four years.[21]

URSUS VANQUISHED

When Kepler had first arrived in Prague, in January 1600, and before meeting Tycho, he had managed to engineer an encounter with Tycho's archenemy, Ursus. Prior to revealing his identity, he had told Ursus that he didn't like the way Ursus had used his (Kepler's) letter to try to boost his (Ursus's) reputation, then hurriedly departed.[22]

While Kepler had been back in Graz, Ursus had died. But Tycho was still determined to have his revenge. So one of Kepler's first tasks on his return to Prague (and one he deeply resented) was to write a lengthy defense of Tycho against the attacks from Ursus. Its importance now lies not so much in that defense as in its contribution to the debate between scientific realists and instrumentalists.

A scientific realist sees science as seeking after the truth. In this view, good scientific theories are either actually true or are reasonable approximations to the truth. Later theories will be better approximations than those they have replaced. Copernicus, for example, genuinely believed that his Sun-centered system was literally true.

Andreas Osiander, on the other hand, was an instrumentalist. In his infamous preface to Copernicus's book, he held that it was not possible to establish the true cause of anything, and therefore that the sole task of an astronomer was merely to provide a reliable basis for calculating planetary positions. The question of truth didn't arise—what mattered was whether the theory could make accurate predictions. Hypotheses need not be true, or even probable, simply useful. (Something of this view can be found in modern interpretations of quantum mechanics.)

Ursus was also an instrumentalist (at least for other people's theories[23]—somewhat confusingly, he turned into a realist when it came to his own theories[24]). Kepler (and probably all astronomers after him) was very much a scientific realist, at least as far as astronomy was concerned. If he had not been a realist, he would never have come up with his laws of planetary motion.

He advanced a number of arguments in favor of the realist position, tearing Ursus's arguments to shreds in the process. For example, he pointed out that although two different hypotheses may make the same predictions for much of the time, it is bound to be the case that they will eventually make different predictions about something.[25] When they do so, one of them can thereby be shown to be false.

He also repeated the argument he had used in *Mysterium Cosmographicum* that, because Copernicus's hypothesis had explanatory power, it was to be preferred over one that merely described.[26]

Kepler did accept that after Copernicus had put forward his Sun-centered system, Tycho was not the only person to see the advantage of a compromise system. But he argued that Tycho's realization that the planets were not carried around by solid crystalline orbs freed Tycho to construct a system that bore a greater resemblance to reality than anybody else's compromise.[27]

GRAZ

In the twenty months following his first meeting with Tycho, Kepler spent only a little over half of this time actually working for him in Prague. In April 1601, he yet again set off for Graz (leaving his wife in Prague) and did not return until the end of August.[28] His father-in-law had died, and he needed to sort out his affairs.[29]

EMPEROR RUDOLPH

Shortly after Kepler's return to Prague, Tycho introduced him to the Holy Roman Emperor Rudolph II. Rudolph was an eccentric and not a very capable ruler. It was said of him by the Habsburg archdukes that Rudolph was only interested in wizards, alchemists, cabbalists, and the like.[30] The Englishman Sir Philip Sidney described him as "a man of few words, sullen of disposition, very secretive and resolute, and extremely spanulated [stiff and formal]."[31] He had a magnificent collection of animals, plants, and minerals. His collection was said to include the horn of a unicorn and some phoenix feathers.[32] He took a delight in pursuing what we would regard today as thoroughly unscientific interests, such as alchemy, astrology, and occult studies in general.

But Rudolph also gathered around himself some of the brightest minds of the age. These included Giuseppe Arcimboldo (who painted the famous portrait of Rudolph made up of various fruits and vegetables), Cornelius Drebbel (who built the first working submarine), and Giordano Bruno (one of the first Copernicans), as well as Tycho and Kepler.[33] He was, at least nominally, and like all the Habsburgs, a Catholic. But unlike his cousin, Archduke Ferdinand of Styria, Rudolph in his more mature years was not concerned about the religious affiliations of those who worked for him. Rudolph's attitude was the main reason why Kepler's time in Prague was the only period in his adult life when he never suffered any religious persecution.[34]

Three people were needed to enable Kepler to arrive at his laws of planetary motion: Kepler himself, Tycho Brahe (together with his treasure trove of extremely accurate planetary observations), and the Emperor Rudolph. Tycho and Kepler would never have found themselves in the same place were it not for Rudolph's patronage, and it was Rudolph's patronage that was to keep Kepler in Prague after Tycho's death. Rudolph was to supply Kepler with status, with an income (at least some of the time), with freedom from religious persecution, and above all with the time to discover his first two laws of planetary motion. We owe Rudolph a greater debt than is generally realized.

At Tycho's suggestion, Rudolph instructed Kepler that he was to collaborate with Tycho in the production of new—and, it was hoped, much more accurate—tables to enable the forecasting of future planetary positions, to be known as the *Rudolphine Tables*. (The title had been ingratiatingly suggested to Rudolph by Tycho.[35]) This would occupy Kepler—intermittently—for the next twenty-five years.

TYCHO'S DEATH

There is what can be thought of as a major design fault in the male urinary system. The problem (named benign prostatic hyperplasia) lies in the collapsible tube (the urethra) that passes through the middle of an organ that expands with age (the prostate). It would be difficult to think of a worse arrangement. Its consequence is that a high proportion of men find it increasingly difficult to urinate as they grow older—about half of all men over fifty are affected by the problem to a greater or lesser extent, and the proportion increases steadily with age. If those affected wait too long before urinating, some find that they cannot do so.

This was Tycho Brahe's predicament. On the evening of October 13, 1601, he set out for a dinner engagement with the Baron of Rosen-

berg. In spite of his deep belief in astrology, he had no idea of the pro-
longed and agonizing death that awaited him. Kepler tells us that the
drink flowed freely, and Tycho felt his bladder getting steadily fuller,
but good manners prevented him from leaving the dinner table to
relieve himself. By the time he got back home, he was totally unable
to urinate. He suffered five sleepless days and nights of unbearable
agony, followed by a period of fever and delirium. In his final hours,
he constantly repeated the words: "Do not let me be seen to have
lived in vain," and he begged Kepler to make use of his Tychonic planetary system in his future work.[36] He died on October 24, 1601, a few
weeks before his fifty-fifth birthday. Kepler sadly records that this
brought an end to thirty-eight years of observational work.[37] Two days
after Tycho's death (and just two months before Kepler's thirtieth
birthday), Kepler was informed that he was to become the new Imperial Mathematician.[38]

Tycho was buried amid much pomp and splendor in Tyn Church
(image 3.2), in the center of Prague, on November 4. The funeral
address was given by the same Johannes Jessenius who had acted as
a neutral observer in Kepler's earlier negotiations with Tycho. Kepler
also composed a eulogy, which was later published. Tycho's tomb
(above which is a wooden carving of him) can still be seen close to the
altar rail in the front of the church.

Everybody loves a good conspiracy theory, and ever since Tycho's
death, conspiracy theorists have enjoyed trying to argue that Tycho
was murdered. Some have even suggested that Kepler was the murderer because he had a strong motive for killing Tycho. This—it has
been argued—was the only way he would ever get at Tycho's valuable
collection of observations. These theories have blossomed, ever since
traces of mercury were found in Tycho's hair. To a conspiracy theorist,
this is clear evidence of murder.[39]

But the idea is ridiculous, on a number of grounds. First, Tycho was
an enthusiastic alchemist, who would therefore have been exposed
to mercury. Second, he also believed in the curative properties of
mercury. So he would undoubtedly have absorbed some mercury into
his body from time to time, either because of accidental exposure or
deliberately. (He could even have taken some mercury in his final days

because he hoped it would alleviate his bladder problem.) Once in the body, mercury doesn't leave it. It acts as a cumulative poison—it finally kills you only when you have taken a sufficient amount of it. So third, if Tycho had really been murdered, the murderer would surely have found a faster-acting poison than mercury.

Image 3.2. The spires of Tyn Church, in Prague.

Additional circumstantial evidence that the mercury in Tycho was self-administered lies in Kepler's statement (over a year before Tycho's death) that his intellect was weakening because of old age[40]—the real reason for this weakening intellect might well have been the early effect of mercury on Tycho's brain. And the conspiracy theories can now finally be laid to rest by the recent discovery, following the exhumation of Tycho's body in 2010, that there was not enough mercury in his body to kill him.[41] Nor were any traces of any other poisons found. So there is no longer any reason to doubt Kepler's account.

Tycho's premature and unbearably painful and drawn-out death had one hugely significant consequence. It meant that Kepler was able to gain full access to Tycho's huge and unprecedented collection of very accurate observations of the planet Mars over time (although not without problems from Tycho's heirs). This in turn led to the biggest breakthrough in theoretical astronomy since Copernicus—the establishment by Kepler of two simple mathematical laws governing the motion of the planets, set out in his book *Astronomia Nova* (*The New Astronomy*).

CHAPTER 4

PRAGUE (1600–1612)—
THE NEW ASTRONOMY

*A*stronomia Nova—*The New Astronomy*—is one of the most important books ever to have been written in the history of science. Over six hundred pages long, in the scholarly translation by William H. Donahue, it is a formidable achievement. Kepler himself usually referred to the work simply as his "Commentaries on Mars" (wording that appears on the title page), but "The New Astronomy, based upon causes" (also on the title page) far better sums up its revolutionary nature.

It was groundbreaking in two ways. First, it brought into use the totally new idea that there is a physical force emanating from the Sun (as foreshadowed over a decade earlier in his *Mysterium Cosmographicum*) that moves the planets in their orbits, thereby bringing about the critical move of astronomy from mere geometry to physics. And second, it replaced the two-thousand-year-old belief in circular motion with the radical conclusion that the planets moved in ellipses: Kepler's First Law.

After Tycho's death, Kepler saw himself as Tycho's natural successor and as someone who could use Tycho's work to achieve something Tycho himself could never have managed. He compared himself to Ptolemy, who had used the work of Hipparchus (another ancient Greek, who lived in the second century BCE) to construct his theory of planetary motion:

> Tycho did what Hipparchus did. Their work concerns the foundations of the building. Tycho has thus achieved an immense work. But no single man can do everything. A Hipparchus needs a Ptolemy who builds up the theory of the other five planets.[1]

The book came about in spite of the efforts of Tycho's relatives. After his death, Tycho's family inherited ownership of his observations, although it was Kepler who actually took charge of them for as long as he could.

Kepler's chief adversary in Tycho's family was the son-in-law, Frans Gansneb Tengnagel von Campp. Tengnagel had a high opinion of himself and took it upon himself to argue with Kepler over the rights to make use of the observations. Kepler wrote:

> The root of the controversies lies in the bad habits and the suspicion of the family, but also in my own passion and the pleasure I take in teasing others. Thus Tengnagel found no small reason for suspecting me of bad designs. I had in my possession Tycho's observations and refused to return them to his heirs. Yet Franz [Tengnagel] was never satisfied with any of my offers to come to an agreement with him by compromise; but he ... suddenly turned against me with threats ... as if I were a low slave.[2]

and:

> I cannot give you any exact information about the state of affairs among the Tychonians, because Tengnagel keeps me away. He is like a dog in the manger, who does not eat any hay himself but does not let anyone else approach either.[3]

Tycho's observations moved back and forth between Kepler and Tengnagel. At one stage, Kepler had to agree not to publish any of his results without Tengnagel's permission. Finally, a compromise was reached under which Tengnagel agreed to allow publication on the condition that he would write a preface to the book. It was this that led to Tengnagel's fatuous introductory paragraph to *Astronomia Nova*, in which he did little more than make a claim for his own importance and make a plea to the reader—without giving any justification—to accept Tycho's system, rather than Kepler's.[4] Arthur Koestler rightly described this preface as the braying of a pompous ass.[5]

But in spite of all the problems Kepler suffered at the hands of Tycho's relatives, he very loyally reminds readers at every opportunity of his deep obligation to Tycho himself. Undoubtedly, this is largely in order to try to neutralize Tycho's troublesome relatives, but it is also true that much of Kepler's praise of Tycho is genuine and heartfelt. He was always a deeply conscientious man, and he reveals (which he need not have done) that Tycho "on his death bed asked me, whom he knew to be of the Copernican persuasion, that I demonstrate everything in his hypothesis."[6] And in his introduction, he tells us that "of [Tycho], in all fairness, most honest and grateful mention is made, and recognition given, since I build this entire structure from the bottom up upon his work, all the materials being borrowed from him."[7] But there are also places in the book where he very effectively demolishes the Tychonic system.

There has been an extraordinary diversity of views on how exactly Kepler managed to arrive at his first two laws of planetary motion. Isaac Newton dismissively maintained (probably for the purpose of self-aggrandizement) that Kepler had simply guessed that the planets moved in ellipses. Bertrand Russell (without the benefit of an English translation of the book) scathingly wrote that Kepler's discovery was one of the most notable examples of what can be achieved by patience without much in the way of genius.[8] Arthur Koestler, although a great admirer, thought the book was disorganized and that Kepler had arrived at his laws partly because the errors in his calculations had fortuitously canceled out.

In fact, as is now widely recognized, Kepler deserves enormous credit for his achievement. The research of modern authors such as William H. Donahue, [9] Bruce Stephenson,[10] and James R. Voelkel[11] has demonstrated the careful construction of *Astronomia Nova* and the clever way in which Kepler sought to persuade his readers of his case.

It is now clear from the work of Donahue, Stephenson, and Voelkel that Kepler is not necessarily always presenting an accurate historical account of how he arrived at his conclusions but is also writing a didactic work aimed at persuading other astronomers that his conclusions had to be correct. Kepler himself hints at this when he says his book was a question "not only of leading the reader to an understanding of the subject matter in the easiest way, but also, chiefly, of the arguments, meanderings, or even chance occurrences by which I the author first came upon that understanding."[12]

Image 4.1. Kepler lived in a house on this site at 4 Karlova from 1607 until he left Prague in 1612.

The story is a complicated one, but it is an essential part of any biography. It would not be right to relegate Kepler's most important discovery to an appendix. Nevertheless, some readers may wish to skim over some of the technical details that follow.

This was the stage in Kepler's life when he managed to suppress his mystical, numerological, and theological speculations. Instead, when he was forced to concentrate on the raw observational data for a single planet, his mathematical brilliance, his sheer perseverance, and his enormous capacity for hard work, as well as his intuitive physical understanding, all came to the fore to produce what we can now see is a solid work of science.

The contrast with his work in *Mysterium Cosmographicum* could not be greater. In *Mysterium Cosmographicum* he had decided in advance on his theory of the five perfect solids, and then—to put it bluntly—fudged the results (by his choice of which perfect solid should go where) until they matched as well as possible with the observations. The whole of chapter 18 of *Mysterium Cosmographicum* is one long explanation of why the observations that disagree with his theory must be wrong. In his work for *Astronomia Nova*, his approach was the exact opposite—he threw out any ideas that did not line up very precisely with Tycho's observations.

The story began when Tycho assigned the orbit of Mars to Kepler, very shortly after his arrival in Prague. Tycho's motive in doing this was probably to limit Kepler's access to all of Tycho's precious observations. His experience with Ursus must have made him very wary of revealing too much of his data. Kepler, ever the optimist, didn't quite see it in this light:

> But when he saw that I possess a daring mind, he thought the best way to deal with me would be to give me my head, to let me choose the observations of one single planet, Mars. This has taken up all my time and I have not been concerned with observations of another planet.[13]

Kepler initially considered that he would solve the problem of the orbit of Mars in only a week, and he even had a bet with Longomontanus (the most senior of Tycho's assistants at that time) to this effect.[14] In fact, the problem took him five years to solve. Nevertheless, Mars was a fortunate choice. Thanks to Kepler's work, we know that all the planets move around the Sun in ellipses (which can be crudely thought of as squashed circles with very well-defined mathematical properties).[15] But the planetary ellipses differ hardly at all from circles. The degree of departure of an ellipse from a perfect circle is measured by a quantity called the eccentricity. A circle's eccentricity is exactly zero. The bigger the eccentricity, the more squashed is the ellipse. With the exception of Mercury (which is difficult to observe because it is so close to the Sun), the orbit of Mars has the greatest deviation from a circle of all the then known planets, as table 4.1 shows.

Table 4.1. Planetary eccentricities—departures from circular orbits.

Planet	Eccentricity
Mercury	0.206
Venus	0.007
Earth	0.017
Mars	0.093
Jupiter	0.048
Saturn	0.054

So if you can solve the orbit of Mars, you can solve the orbit of any of the other planets. Kepler recognized his good fortune in being given Mars, although he attributed this to God: "I ... think it to have happened by divine arrangement that I arrived at the same time in which [Longomontanus] was intent upon Mars, whose motions provide the only possible access to the hidden secrets of astronomy, without which we would remain forever ignorant of those secrets."[16]

The choice of a single planet was fortunate for a second reason. It forced Kepler away from his obsession with the five perfect solids. It was this obsession that had been his main motivation for coming to work for Tycho in the first place. Instead, by requiring him to channel all his considerable energies on to just one planet, it made him focus on one of the other (and significantly less prominent) themes of *Mysterium Cosmographicum*—the idea that there was a force emanating from the Sun that caused the planets to move in their orbits. This was an idea whose time had finally come. Admittedly it had a distinctly mystical overtone—neither he nor (much later) Isaac Newton was able to explain the mechanics of how this action at a distance could possibly work. Nor would a credible explanation come along until Albert Einstein arrived at his general theory of relativity some three hundred years later. But, thanks to Tycho Brahe's work, the old Aristotelian idea that the planets were moved in their orbits by gigantic crystalline spheres was no longer credible.

Kepler cleverly wrote his book in such a way that anybody who read and fully understood it would be forced to the conclusion that the older theories could not be made to agree with Tycho's very accurate observations, no matter how hard one tried. Something new was needed. This new approach would result in a fundamental reassessment of the Earth's orbit around the Sun and a departure from circular motion for all the planets, including Earth—the ellipse.

The way in which Kepler chose to arrange the material in *Astronomia Nova* so as to convince others of his conclusions can be split up into eight stages.[17] (But it is worth emphasizing again that, as Donahue, Stephenson, and Voelkel have all demonstrated, this did not necessarily correspond to an accurate historical account of how Kepler actually arrived at his conclusions.)

Stage 1: From the perspective of observers on Earth, the Sun seems to move relative to the background of stars along a path called the ecliptic, returning to essentially the same point after one year. But it doesn't quite do so at a uniform rate. It moves at a more rapid rate in January and at a slower rate in July.[18] This can be explained, in part, by assuming that the Sun is offset a little from the center of the Earth's orbit. (What is also happening, and what Kepler was shortly to discover, is that because the Earth is a little closer to the Sun, in January, it moves more quickly. When it is a little farther away from the Sun, in July, it moves more slowly.) So in doing calculations, an astronomer could either introduce a notional "mean Sun," which travels at a constant speed, or make use of the real Sun. Ptolemy, Copernicus, and Tycho had all opted for this mean Sun, but Kepler took the real Sun as his reference point. The mean Sun, after all, has no physical significance. It is little more than a geometrical abstraction, only appropriate if you consider that astronomy is just an exercise in geometry.[19] Kepler often had a finely tuned sense of physical reality. Even when he had written *Mysterium Cosmographicum*, he had already recognized that using the real Sun would be more successful.[20] In his early days at Benatky, he had somehow persuaded Tycho (probably against Tycho's better judgment) that he should be allowed to make all his measurements and calculations relative to the real Sun.[21] And it did indeed prove to be a far more successful approach.

Stage 2: His next critical breakthrough was the realization that the planes of the orbits of all the planets pass through the center of the Sun.[22] In other words, if you draw the orbit of any planet on a flat piece of paper, the center of the Sun will also appear on the same piece of paper—it will not be above or below it.

This is an implicit part of what we now know as Kepler's First Law (and an inevitable consequence of Isaac Newton's law of gravitation, which would not be developed for another seventy years), but it is such an important realization that it is worthy of being a law in its own right. Professor Owen Gingerich has suggested that we could even label it as Kepler's "zero-th" law.[23] As Kepler pointed out, neither Copernicus nor Tycho had appreciated this vital point. And it is certainly evidence that the Sun does somehow control the motion of each planet.

Stage 3 is where Kepler did his demolition job on Ptolemy, Copernicus, and Tycho. When he started out on his search for Mars's orbit, he probably still believed that he would find it to be some form of circular motion. By the end of his search, he had definitively established that this could not possibly be the case.

Ptolemy had thought that the Sun, the Moon, and all the planets moved in independent orbits around the Earth. But he also needed to explain the strange backward motion of the planets that occurred from time to time. To do this, he made use of the earlier Greek idea that the planet itself moved on an epicycle, the center of which (an empty point in space) was on a circular orbit around the Earth. However, he knew that the center of the epicycle didn't quite move at a constant rate. So to make theory correspond with observations, he had to displace the Earth a little from the center of this orbit. He also assumed the existence of a point called the equant, the point from which the center of the planet's epicycle *appeared* to be moving at a constant rate. This elaborate fudge, as shown in figure 4.1, was the only way Ptolemy could see to account for the observations, and it was a departure from the Greek ideal of uniform circular motion.

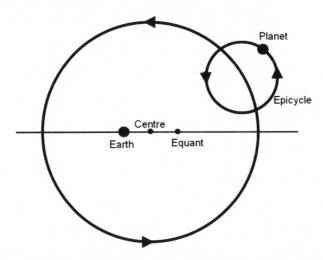

Figure 4.1. Ptolemy's model. Each planet has a complicated motion around the Earth. The center of the epicycle is an empty point in space.

Copernicus's stroke of genius was to realize that the epicycle was merely a reflection of the Earth's own motion around the Sun. So he took the revolutionary step of placing the Sun at the center of the system and placing Earth and the other planets in orbit around the Sun. The question for Kepler then became whether planetary motion could be accounted for in the traditional way, but without the complication of the epicycle (figure 4.2).

The observations showed that in a Sun-centered system, just as in an Earth-centered system, the planets still did not move at a completely uniform rate. So could the motion of a planet be described by a circular motion at a rate that seemed uniform only from an equant point (which was offset from the center), and with the Sun also offset from the center but in the other direction? This was the simplest reasonable set of assumptions. It was analogous to the model shown in figure 4.1 and is as shown in figure 4.2. This was the problem Kepler had to tackle for Mars.

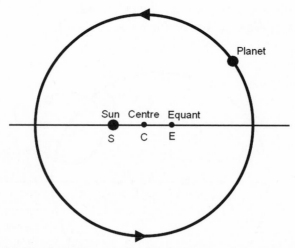

Figure 4.2. Was it possible to describe a planet's orbit in this manner?

The Earth takes just over 365 days to orbit once around the Sun, and Mars takes 687 days.[24] This means that about once every two years (more precisely, every 780 days on average), the Sun, the Earth, and Mars will line up in a straight line. Mars is then in the exact opposite position in the sky to the Sun and is said to be in opposition. So an observation of the position of Mars from Earth (relative to the background stars) at that moment is the same as an observation taken from the Sun—in effect, the Earth has been eliminated from the observation.

On the assumption that an equant point exists, observations taken at opposition can be used to try to determine a circular orbit for Mars, but this can only be done by a lengthy and monotonous trial-and-error method, involving a series of ever-closer approximations. In chapter 16 of his book, Kepler describes how ("by a most laborious method"[25]) he used four[26] of Tycho's observations of Mars[27] at opposition for this purpose. Kepler makes clear to his readers the sheer tedium of this:

> If this wearisome method has filled you with loathing, it should more properly fill you with compassion for me, as I have gone through it at least 70 times at the expense of a great deal of time.[28]

But he finally ended up with a circular orbit that matched all four observations. He then proceeded to check this orbit against a further eight observations of Mars, again at opposition.[29] He found that the orbit matched these observations to an accuracy of two minutes (2/60 of a degree),[30] which was the level of error in Tycho's very accurate observations.

On the face of it, Kepler had triumphed. But the opening to chapter 19 of his book puts paid to this premature sense of victory:

> Who would have thought it possible? This hypothesis, so closely in agreement with the [opposition] observations, is nonetheless false.[31]

The problem that Kepler had discovered was this: A planet moves in a path that never strays far from the ecliptic. Mars, as seen from Earth, takes 780 days to journey around the sky (relative to the background stars) from one opposition to the next, hugging the ecliptic as it travels. But its angular distance from the ecliptic also changes with time, although not by very much—just a few degrees either side of the ecliptic. Kepler had cleverly used two of Tycho's observations of Mars's distance from the ecliptic[32] to perform an independent check on his first set of calculations.

Unfortunately, they gave different values for the parameters of the orbit. If he tried to combine the two different results by dropping his assumption that the distances SC and CE (in figure 4.2) were different, then his orbit no longer matched the observations at opposition. At worst, they were off by a full 8 minutes (8/60 of a degree).

Before Tycho Brahe's very accurate measurements, such a difference would have been lost in the measurement error.[33] But Tycho's observations were accurate to an unprecedented level of a mere 2 minutes, and Kepler had too much respect for Tycho's abilities as an observer to ignore this simple fact. Anybody else (and Kepler himself in his earlier years) would probably simply have added some kind of fudge factor and declared the problem solved. But the more mature Kepler realized that this was an error that simply could not be neglected. Something was wrong with the basic assumptions:

> Therefore, something among those things we have assumed must be false. But what was assumed was that the orbit upon which the

planet moves is a perfect circle; and that there exists some unique point [i.e., the equant] . . . about which Mars describes equal angles in equal times. Therefore, of these, one or the other (or perhaps both) are false, for the observations used are not false.[34]

Either the assumption of circular motion or the assumption of the equant (or both) was wrong. He saw that he would have to throw out one or both of these assumptions of his predecessors and start all over again. As he put it, in perhaps the most significant sentence in the whole book:

Now, because they could not have been ignored, these 8 minutes alone will have led the way to the reformation of all of astronomy.[35]

Kepler has actually been very cunning in writing this section of his book. He has persuaded the perceptive reader that, by dint of his hard work, he has demonstrated that none of the old models fit Tycho's very accurate observations, so all must be false. Something new was needed. And the reader is forced to agree that this is necessary.

Stage 4: He now recognized that he was going to have to throw out one or both of the assumptions that had been at the core of astronomical thinking for the previous two thousand years. But first, and more fundamentally, he was going to have to check the Earth's orbit—if the Earth did not move at a uniform rate around the center of its orbit (which was also a key assumption of Copernicus, and in effect of Ptolemy and Tycho as well), then observations made from Earth that made use of this assumption would be wrong. Kepler had always wondered why the Earth should be different from the other planets in this respect.[36] Now he was going to have to find out whether it really was.

But how do you find out whether the Earth moves at a uniform rate? Kepler's solution was, in the words of Albert Einstein, "an idea of true genius."[37] He measured the Earth's orbit as if from a fixed platform in space. To do this, he made use of Mars—the very planet whose orbit he was trying to determine. He knew that every 687 days, Mars returned to the same spot in its orbit, at the same distance from the Sun, and in the same position against the background of the celestial sphere (as seen from the Sun). So he noted four positions of Mars relative to Earth (from which he could in turn work out the position of Earth relative to Mars) at intervals of 687 days.[38] This succession of Tycho's observations at 687-day intervals enabled Kepler to plot the true position of Earth at various times in its orbit. From this he was able to conclude that the Earth does not revolve around the center of its orbit at a uniform rate after all:

> It has certainly been demonstrated that [the Earth] is moved in a non-uniform manner, that is, slowly when it is farther from [the Sun] and more swiftly when it has approached [the Sun].[39]

Like the other planets, the Earth, too, seemed to have something like an equant point, around which it appeared to move uniformly. This was an extremely important finding, because it meant that the Earth behaves in just the same way as the other planets. (William H. Donahue points out that it was actually this discovery that did more to improve the accuracy of planetary forecasts than either of his first two laws of planetary motion.[40])

Stage 5: In chapter 32, Kepler notes that all the planets (now—thanks to his earlier demonstration—including even the Earth) move faster when they are closer to the Sun and slower when they are farther away. At the closest and farthest points, Kepler shows that the speed is inversely proportional to the distance from the Sun and that, at these points, this is mathematically equivalent to Ptolemy's idea of an equant. But (as foreshadowed in his *Mysterium Cosmographicum* some years earlier) he then argues that, rather than using the equant (which is a mere mathematical abstraction), it makes more sense to ascribe this change in speed with distance to a physical force coming from the Sun. The force is stronger when the planet is closer to the Sun (which is why the planet then moves more quickly) and weaker when it is farther away (so the planet then moves more slowly). This gives what is only the illusory appearance of an equant point.

It turns out that the equant introduced by Ptolemy (and which Copernicus had hated and had tried to get rid of) is actually just a good geometrical approximation to the physical reality—created by physical causes—that a planet moves faster when closer to the Sun and slower when farther away. The Sun occupies one of the ellipse's two foci, and the empty focus is equivalent to the equant point; to a good first approximation, the planet moves at a constant angular rate when seen from this empty focus. This was one of Kepler's most significant discoveries.[41]

There is, he realizes, just one argument against his position. If all the planets—including Earth—are indeed moved by the Sun, and the Earth does not move the Sun, how is it that the Earth so obviously moves the Moon? If the Earth can move the Moon, why can't it—as Tycho maintained—move the Sun as well? To get around this problem, Kepler asserts that the Moon is "akin to the Earth in its corporeal disposition."[42] If the two bodies are alike in some way, then perhaps it is to be expected that they will attract each other.

This may be the reason why Kepler seems to be anticipating Newton's law of universal gravitation. He expresses this most clearly in a letter written at about the same time as *Astronomia Nova*:

> If one would place a stone behind the Earth and would assume that both are free from any other motion, then not only would the stone hurry to the Earth, but also the Earth would hurry to the stone; they would divide the space lying between in inverse proportion to their weights.[43]

and in the book itself:

> It would be preferable to attribute to the Earth a force that retains the Moon, like a sort of chain, which would be there even if the Moon did not circle the Earth at all. . . . The Moon is not driven primarily by the Sun, in its circling of the Earth, but by a power lying hidden in the Earth itself.[44]

and

> If the Moon and the Earth were not each held back in its own circuit by an animate force or something else equivalent to it, the Earth would ascend towards the Moon by one fifty fourth part of the interval [between them] and the Moon would descend towards the Earth about fifty three parts of the interval, and there they would be joined together.[45]

These arguments are there as a result of the need to justify the otherwise apparently anomalous position of the Earth/Moon system, rather than an anticipation of Newton. Kepler certainly had a theory of gravity, but it fell short of Newton's *universal* theory of gravity. Tantalizingly, though, he did recognize that

> it follows that if the Moon's power of attraction extends to the Earth, the Earth's power of attraction will be much more likely to extend to the Moon and far beyond.[46]

It was these ideas that enabled him to be one of the first people to realize that the Moon has an attractive power over the Earth:

> If the Earth should cease to attract its waters to itself, all the sea water would be lifted up and would flow on to the body of the Moon. The sphere of influence of the attractive power in the Moon is extended all the way to the Earth.[47]

For centuries, sailors had recognized that there was a relationship between the tides and the position of the Moon in the sky. But Kepler was one of the first people to realize that the tides are actually caused by the attraction of the Moon. Even before he wrote *Astronomia Nova*, in March 1598 he had made this suggestion to his friend Herwart von Hohenburg.[48] But just as the planets revolved around the Sun only because (according to Kepler) the Sun rotated, the only reason the Moon revolved around the Earth was because the Earth rotated. In this, he was wrong.

Stage 6: Kepler now tries to demonstrate what the nature of the force from the Sun consists of. He was absolutely right that the Sun exerts a force on the planets, and that this is what keeps them in orbit. But he was completely wrong on the physics of how the force operated, and his explanation is now of no more than historical interest. Stage 6 should therefore not now be taken at all seriously. He correctly suggests that the motive power coming from the Sun was something like light, which also weakens with distance. He says that this solar emanation:

as a consequence of the rotation of the Sun, also rotates like a very rapid whirlpool throughout the whole breadth of the [Solar System], and carries the planets along with itself.[49]

He draws an analogy between the Sun's action on the planets and an orator addressing a huge crowd of people who surround him on all sides. As the orator turns around, everyone in the crowd in turn gets a glimpse of the orator's eyes. Using this analogy, Kepler concludes that in order to cause the planets to move, the Sun must rotate.[50]

In this, he was incorrect. He thought the natural state of a body such as a planet was to be stationary.[51] So something was needed to give a constant push to each planet. He didn't know what had been discovered by his contemporary Galileo (and what was later incorporated into Newton's first law of motion)—that a moving body will continue to move in a straight line at a constant speed unless acted on by an external force (such as friction or gravity). Kepler would have argued that if the Sun were magically to disappear, the planets would simply stop moving. In fact, if the Sun disappeared, they would continue to move but would instead move in straight lines.

Nevertheless, this was yet another occasion when he was partially right, but for completely the wrong reasons. He was right that the Sun does rotate. (It takes about a month to do so. As it isn't a solid body, different latitudes rotate at slightly different rates.) But he was totally wrong in his fundamental idea that its rotation is the cause of the planets' motion. Kepler came to the conclusion that the Sun rotates (and he was the first person ever to do so) only a few years before the discovery of sunspots moving across the Sun, and the consequent realization that the Sun must indeed rotate, carrying the sunspots around as it did so. He saw the discovery of this rotation as confirmation of his idea[52] that the rotation caused the planets to move, although in reality it was not.

He picks up on the work of William Gilbert, the English physician who had published a work on magnetism a few years earlier. Gilbert had demonstrated that the Earth is a huge magnet. Could magnetism, or something closely analogous to it, be the cause of planetary motion? Kepler certainly thinks so—the Sun, too, must be magnetic, and the rotation of the Sun caused the Sun's magnetic motive force to rotate and to drag the planets around with it. (He was again partially right here, but again for the wrong reason—the Sun does have a magnetic field, but this has nothing to do with the movement of the planets around the Sun.)

This still left a problem. Each planet varied its distance from the Sun. What was the mechanism for this? He asks us to imagine a cable suspended over a river and joining the two banks. He then draws an analogy between a planet and a ferryman. The ferryman uses a combination of his rudder and the flow of the river to move his boat back and forth from bank to bank, at right angles to the flow of the water.[53] Just as the ferryman can move back and forth, so a planet can move itself toward and away from the Sun. But how did a planet know what its distance from the Sun was? The most likely possibility seemed to be that each planet "observes the increasing and decreasing size of the solar diameter and understands (using this as an indication) what distances from the Sun it should arrive at any given time."[54] (Kepler had already succeeded in measuring the tiny difference in the apparent diameter of the Sun when the Earth was at its closest and farthest from the Sun. He had described the apparatus he had used to do this and the results he had obtained in great detail in his *Optics*.[55] If he could make this measurement, then no doubt the planet could also make it!)

Stage 7: Kepler had already noted that the speeds of a planet at its maximum and minimum distances from the Sun were inversely proportional to those distances, and that this was a relationship that was roughly true over the whole of the planet's orbit.[56] He now wanted to use this information as a way of calculating where a planet would be in its orbit at any particular time.

The first step was to recognize that the time taken by a planet to move a tiny distance around its orbit was proportional to its distance from the Sun at that point. So he reasons that if he were to add up all these distances from the Sun, he would have a measure of the time taken to get from one point on the orbit to another. But this was a tricky calculation. He starts by dividing up the orbit (still assumed at this point to be circular) into 360 triangular segments with their vertices at the Sun (figure 4.3).[57] This was a cumbersome procedure. Was there an easier way? He eventually realizes that he could use the areas of the 360 triangles he had created as a stand-in for the sums of the distances. "It therefore seemed to me I could conclude that by computing the area CAH or CAE, I would have the sum of the infinite distances in CH or CE."[58]

So in chapter 40, he eventually arrives at the conclusion that

> Therefore ... as the area CDE is to half the periodic time, which we have proclaimed to be 180°, so are the areas CAG, CAH to the elapsed times on CG and CH. Thus the area CGA becomes a measure of the time ... corresponding to CG.[59]

or, as we would put it today, the Earth and the other planets sweep out equal areas in equal times (figure 4.4). This is now known as Kepler's Second Law, which he actually discovered before his First Law. At this stage, though, he saw this just as an approximation to his (ultimately incorrect) distance law, rather than the exact law that he later realized it to be.

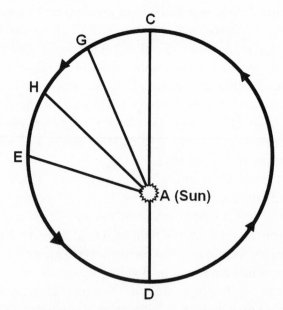

Figure 4.3. A simplified version of Kepler's diagram illustrating how he arrived at his Second Law. Only 3 of his 360 segments are shown.

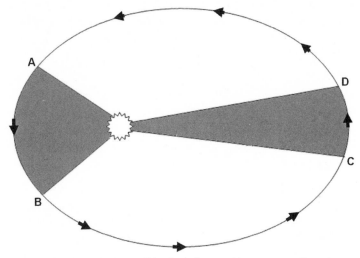

Figure 4.4. The final version of Kepler's Second Law—as a planet moves in its elliptical orbit from A to B, it moves at a greater speed (because it is closer to the Sun) than it does along the shorter distance from C to D. The result is that if the gray area bounded by A, B, and the Sun is the same as the gray area bounded by C, D, and the Sun, then the time taken to travel from A to B is the same as the time to travel the shorter distance from C to D.

So if the shape of the orbit is known, and the time taken to complete one circuit of the orbit is also known, then the Second Law enables you to compute the position of a planet in its orbit at any time in the future from a given starting position.

Stage 8: Throughout the book, Kepler drops occasional hints[60] that his eventual conclusion will be that the planets' orbits are not after all circular. He seems to have realized this quite early on, but it took a considerable time—and several chapters of his book—before he appreciated that these noncircular orbits were ellipses. To a modern reader (because we know his eventual conclusion), his explanations seem painfully and unnecessarily lengthy and drawn-out. Initially, he is only able to demonstrate that Mars's orbit was some sort of oval—the word is derived from the Latin word *ovum*, meaning an egg—and an oval, or egg shape, does not necessarily have the clearly defined mathematical features of a circle or an ellipse. (A circle or an ellipse will always have two axes of symmetry, but an oval may only have one, making it mathematically much more problematic.)

Clearly then … the orbit of the planet is not a circle, but comes in gradually on both sides and returns again to the circle's distance at perigee. They are accustomed to call the shape of this sort of path "oval."[61]

However, the realization that the planets departed from circular motion must have been a blow. Circles were easy to perform calculations on. They were a natural—indeed perfect—shape. They had been used in astronomy for the previous two thousand years. How could anybody even think of doing calculations with an oval instead? But he struggled on. In July 1603, he wrote to his friend David Fabricius that

if only the orbit were a perfect ellipse, all the answers could be found in the work of Archimedes & Apollonius.[62]

Nevertheless, he proceeds to use an ellipse as an approximation to the oval orbit he was trying to pin down. It is only later, in chapter 58, that he finally realizes that his carefully calculated oval falls halfway between the circle and the approximating ellipse. And the only shape that can do this exactly is also an ellipse:

But the only figure occupying the middle between a circle and an ellipse is another ellipse. Therefore the ellipse is the path of the planet.[63]

Kepler's First Law—that the planets move around the Sun in ellipses—had finally emerged. However, as Donahue points out, nowhere in *Astronomia Nova* does Kepler actually state the other half of the law, that the Sun was at one focus of the ellipse.[64] This would have to wait for later.

For the moment, though, this was not important. This time, Kepler's triumph was genuine. With his recast orbit for Earth and his two laws (the First Law describing the shape of the orbit, and the Second Law locating the position of the planet on that orbit at any instant of time), he now had the means to forecast planetary positions to a much higher level of accuracy than ever before in the history of astronomy. Almost as an aside, in the introduction to *Astronomia Nova*, he notes that hitherto, it has not been possible to do this with sufficient certainty. In fact, in August 1608, Mars was a little less than 4° beyond the position given by calculation from the *Prutenic Tables*.[65] In August & September 1593, this error was a little less than 5°, while in my new calculation the error is entirely suppressed.[66]

All that remained for him to do was to produce his own set of tables—the *Rudolphine Tables*, named after his patron Rudolph II—based on his discoveries, which would enable the accurate prediction of future planetary positions for the first time in human history. But this work would not be completed until eighteen years after the publication of *Astronomia Nova*, and long after Rudolph's death.

KEPLER'S INTRODUCTION TO *ASTRONOMIA NOVA*

When *Mysterium Cosmographicum* had been published, over ten years previously, the religious establishment at Tübingen University had refused Kepler permission to include his attempt to reconcile biblical teaching with Copernicanism. Now that he was his own master, Kepler was at last able to speak his mind on this topic. It is worth outlining Kepler's views, which take up a significant part of his introduction to *Astronomia Nova*, particularly as this introduction was the only one of his writings to appear in English before the late nineteenth century.

One of the teachings of Jesus, to be found in three of the four Gospels in the New Testament, is that—although all other sins could in principle be forgiven—a slander against the Holy Spirit (the third person in the Christian Trinity) could never be forgiven by God.[67] Unfortunately, Jesus neglected to clarify what exactly constituted a slander against the Holy Spirit. This regrettable omission left generations of pious Christians in fear of spending an eternity of suffering in hell for having accidentally committed this particular sin.

One common belief was that the sin included disagreement with any text in the Bible, since it was thought that the whole of the Bible had been dictated by the Holy Spirit. Copernicanism was in conflict with several biblical passages that proclaimed that the Earth was fixed[68] and was therefore committing this very sin. Kepler makes clear that he is fully aware of this argument and also that he realizes that this was the real reason why most people felt unable to accept Copernicanism:

> There are, however, many more people who are moved by piety to withhold assent from Copernicus, fearing that falsehood might be charged against the Holy Spirit speaking in the scriptures if we say that the Earth is moved and the Sun stands still.[69]

He counters it by explicitly rejecting the belief in literal inerrancy. His argument is that we often speak as though something were literally true ("We are carried from the port, and the land and cities recede."[70]) because it is a convenient way of expressing ourselves, even though we know that what we are saying is not actually the case. The Bible, he argues, is no different when talking about the Earth being fixed. (It was, though, unfortunate that the authors of these verses had said that the Earth *was* fixed, rather than merely *seemed to be* fixed.) He proceeds to list examples from the Bible that he thought could not possibly be taken literally[71] and demonstrates—at least to his own satisfaction— that he can reinterpret the biblical passages that entail a fixed Earth.[72]

It was a clever argument, and one that was perfectly logical in the context of the times. But he could not have foreseen that it was just the first phase of a dangerous and slippery slope for religious belief, in which our increasing scientific and other knowledge over the following centuries would mean that many more biblical passages could no longer be taken as literally true.

The introduction to *Astronomia Nova* is also where he demolishes Tycho's compromise system most effectively. Tycho held that the Sun moved the other five planets, but that the Earth moved the Sun. Kepler points out how unlikely this is. It is, he says, just too much of a coincidence that—if Tycho is right—the distance between the Sun and the Earth, and the Sun's time to go around the Earth are *both* intermediate between those of Venus and Mars:

> What shall I say of the [Earth's] periodic time of 365 days, intermediate in quantity between the periodic time of Mars of 687 days and that of Venus of 225 days? Does not the nature of things cry out with a great voice that the circuit in which these 365 days are used up also occupies

a place intermediate between those of Mars and Venus about the Sun, and thus itself also encircles the Sun, and hence that this circuit is a circuit of the Earth about the Sun and not of the Sun about the Earth?[73]

PUBLICATION

In spite of the difficulties created by Tengnagel as well as the emperor's initial failure to come up with the funds for printing the book, it was eventually published in 1609, four years after Kepler's triumphant discovery of planetary elliptical motion.[74] The title page proclaimed that the book was "by order and munificence of Rudolph II," as well as making clear that Kepler's results were derived "from the observations of Tycho Brahe."

In a lengthy dedication to the Emperor Rudolph II, he describes his five-year struggle with the planet Mars as a war in which he has finally triumphed and has taken Mars captive, "since for some time he has been accustomed to dropping his vaulted shield and his arms, & giving himself over freely and playfully to capture and bondage."[75] He lists all the defeats and disasters that have occurred in his long military campaign—as always, he is unable to resist a reference to his money problems, by putting the "extreme deficiency of provisions"[76] at the top of the list. And finally:

> When he [Mars] saw that I held fast to my goal, while there was no place in the circuit of his kingdom where he was safe or secure, the enemy turned his attention to plans for peace: sending off his parent Nature, he offered to allow me the victory; and, having bargained for liberty within limits subject to negotiation, he shortly thereafter moved over most agreeably into my camp.[77]

Kepler had decisively won the war.

CHAPTER 5

PRAGUE—MANY NEW THINGS

"I start many new things, even though the earlier thing is still unfinished."[1] Kepler was incapable of concentrating on one thing, and seeing it through to completion, to the exclusion of all other activities. His hyperactive personality meant that, while toiling away on *Astronomia Nova*, he also poured out works on a whole host of other subjects. Many of these were entirely original in their content. Above all, they demonstrated his insatiable curiosity over a wide range of topics. In addition, he carried on an active correspondence with other researchers. His time in Prague was the most productive period of his life.

OPTICS

By far the most important of these other works was his book on optics, on the study of which Kepler was a pioneer. He interrupted his work on the orbit of Mars, in the year 1603, to work on a lengthy treatise on the subject. The English translation (by William H. Donahue) runs to over four hundred pages. As Donahue has put it, "The result is one of the most important optical works ever written which, even when it is wrong, is wrong in an interesting and fruitful way."[2] In effect, Kepler founded the modern science of optics.

One of the main topics in the book was that of atmospheric refraction. Astronomers curse the Earth's atmosphere. Atmospheric turbulence weakens and distorts the light coming to us from the stars—it is this turbulence that makes the stars appear to twinkle. The atmosphere's other pernicious effect is to bend (refract) the light from any heavenly body as it passes through lower and increasingly dense layers of air. This makes it seem to come from a slightly different direction (see figure 5.1), higher up in the sky—the same phenomenon that makes a straight stick appear bent if it is half submerged in water. So something above the Earth's atmosphere appears very slightly higher in the sky than it really is.

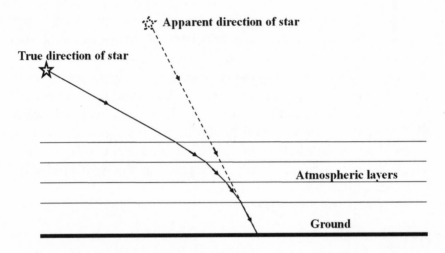

Figure 5.1. As light from a star passes through lower and steadily denser layers of atmosphere, it is deflected more and more from its original path.

Light from a star in the zenith (the point immediately overhead in the sky) doesn't suffer any refraction. Even halfway from the zenith to the horizon (at an angle of 45° from the horizon), the refraction amounts to less than 1 minute (1/60 of a degree)—the amount of refraction has been exaggerated in figure 5.1 to illustrate the phenomenon. However, at about 20° above the horizon, the refraction starts becoming significant, amounting to about 2.5 minutes. Below 20°, the amount of refraction increases rapidly. At the horizon, light is refracted by just over half a degree. (Coincidentally, the apparent diameter of the Sun is about half a degree, so refraction enables us to continue to see the Sun for a few minutes after it has technically set.)

If an astronomer made no correction for this effect on stars and planets, particularly when they are at low altitude, their positions would be incorrectly recorded. Tycho Brahe had been well aware of the problem and had drawn up a table of corrections to be applied to his observations. These were close to the values we use today. Kepler wrote extensively on the subject, providing a theoretical justification for Tycho's corrections.

Kepler did not manage to discover the precise mathematical law governing the refraction of light, although three other people in his time did so independently. (The Arabs had also discovered this law several hundred years previously.) It is now known as Snell's Law, but in fact it was first discovered in the West in 1602 by the English astronomer and mathematician Thomas Harriot, who unfortunately failed to publish his findings. Harriot and Kepler corresponded briefly on the topic of refraction in 1606, but Harriot never mentioned the law to Kepler. It was then independently rediscovered by the Dutch astronomer and mathematician Willebrord Snellius (after whom it is named) in 1621, and also by the philosopher and mathematician René Descartes a few years later. However, Kepler did make the important discovery that the intensity of a light source declines in proportion to the square of the distance it has traveled[3]—what we now know as the inverse square law.

Kepler also examined the structure of the human eye. For this, he drew on the work of physicians and anatomists of the time, including his good friend Johannes Jessenius. But he took this work further and described the mechanisms by which we see, "which no one at all to my knowledge has yet examined and understood in such detail."[4] He accurately describes how rays of light from any point outside the eye reach the eye in a cone of radiation, are refracted by the lens of the eye, and converge on a single point on the retina, where an image is then formed. He is the first person to acknowledge that an object would form an inverted image on the eye's retina, although he fails—unsurprisingly—to come up with a good explanation of how the brain manages to correct for this. To justify his conclusion that the image is indeed inverted, he draws the analogy of a spherical bowl of water that, when placed near a window, forms a clear inverted image of the window a semi-diameter away from the bowl.[5] He also takes the opportunity to reject decisively the idea that objects were illuminated by rays emitted by the eye, an idea others had put forward.[6]

Spectacles were already in use in Kepler's time. As he put it, "Those who see distant things distinctly, nearby things confusedly, are helped by convex lenses. On the other hand, those who see distant things confusedly, nearby things distinctly, are aided by concave lenses."[7] But he was the first person ever to explain why this was so. Having correctly dismissed other explanations as just plain wrong, he points out that people with long sight, "when they use convex glasses, they alter the cone of radiation of a nearby point so as to [make it] appear to arrive as if from a distance,"[8] and vice versa for shortsighted people using concave lenses (see figure 5.2).

He spices up an otherwise dry and lengthy account on optics with occasional personal anecdotes:

I know two men ... one of whom reads the tiniest fine print, but brings his eye so close that he cannot use both his eyes at once. The same man cannot reach with clear vision as far as ten paces, beholding nothing but clouds. Nevertheless, lenses of deep concavity help him to see more distant things, although these same lenses completely confuse my vision, even though I also use ones that are concave but more moderate. The other man ... was nearly blind for nearby things, but was [exceptionally clear-sighted] for distant things, so much so that he prided himself in being able to count, on a house at a [great] distance, the new roof tiles mixed in with the old. Using convex glasses and a paper unfolded at a distance, ... he used to read not badly.[9]

It is from this passage that we know that Kepler himself wore spectacles to correct his own short sight, although they are absent in all the portraits of him.

Kepler reveals his other eye defect when he gives an explanation of astigmatism and says, "Hence it is that those who labor under this defect see a doubled or tripled object in place of a single narrow and far distant one. Hence in place of a single moon, ten or more are presented to me."[10] His condition was almost certainly a consequence of the smallpox he caught when he was three years old.

Kepler also devotes a large section of the book to the Moon. Although it was written some seven years before Galileo turned his telescope on the Moon, Kepler makes clear that he already thought of the Moon as a body not unlike the Earth. He concludes from the roughness of the dividing line between the areas of the Moon in sunlight and those in darkness that some parts of it must be low and others high.[11] He knew that he was in agreement with Plutarch on this, and that the Aristotelians—who held that the Moon was a perfect body—would disagree.

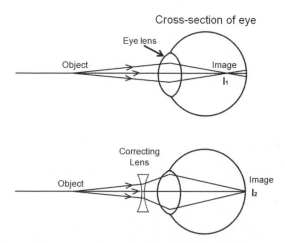

Figure 5.2. The diagrams show a cross-section of an eye and how light passes through it to the retina, at the back of the eye. In a short-sighted person (such as Kepler), light from a distant object comes to a focus in front of the retina, at I_1 (upper diagram) so the object is out of focus at the retina. Kepler was the first person to explain that a concave lens placed in front of the eye (lower diagram) makes the light seem to come from a nearer point, so that it comes to a focus at I_2 on the retina.

When the Moon is seen as a narrow crescent, it is possible to make out the rest of the Moon's surface, even though it is not receiving any direct sunlight. (This is sometimes known as the old Moon in the new Moon's arms.) There had been a number of explanations put forward for this. Some thought that the Moon had a faint light of its own, some thought that a tiny amount of sunlight passed right through the Moon and illuminated its other side. Tycho Brahe had even suggested that the illumination was caused by light from Venus. But Kepler repeats the correct explanation, first put forward by his teacher Michael Maestlin, that the dark surface is illuminated by reflected sunlight from Earth.[12]

He also gives the first essentially correct explanation of why the Moon becomes a reddish color when it is being eclipsed,[13] just as the eight-year-old Kepler had first observed in Leonberg many years previously (and as described in chapter 1).

ASTROLOGY

Any admirer of Kepler has to acknowledge the unfortunate fact that he had a deep and enduring belief in astrology, and that this was an integral part of his worldview. He was the last astronomer of any significance ever to have such beliefs. He once famously described himself as "a Lutheran astrologer."[14] In his defense, it can be said that he was in this respect little more than a man of his times. In a climate in which almost everyone else (including, for example, Tycho Brahe and Rudolph II) believed in astrology, his views were perfectly understandable.

However (exactly as with his Lutheranism), he did question more orthodox views on the subject and arrived at his own unique and carefully thought-out position. But he never saw the need to discard the belief altogether. Even though a few people were starting to express skepticism about astrology, he warned them not to throw out the baby with the bathwater.[15]

He once summarized his view of astrology in a clever analogy that he set out in a letter of 1599 to his friend Herwart von Hohenburg:

> How does the face of the sky affect the character of man at the moment of his birth? It affects the human being—for as long as he lives—in the same way as the knots which the peasant haphazardly puts around the pumpkin. They do not make the pumpkin grow, but they decide its shape. So does the sky: it does not give the human being morals, happiness, children, fortune and wife, but it shapes everything in which the human being is engaged.[16]

In other words, he saw astrology in much the same way as we now see our genes—not as a complete determinant of all our actions but as a strong and inescapable influence on us.

In the same letter, he went on to describe how the positions of the planets at her birth had affected the character of his wife, Barbara. Already, after only two years of married life, their relationship had become problematic, but at least it could be blamed on the stars:

> Look at the human being at whose birth the constellation of Jupiter and Venus were not fortunate. You will see that such a human being can be just and wise, but has a less happy and rather sad fate. Such a woman is known to me. She is praised throughout the city on account of virtue, chastity and modesty. But she is simple-minded and stout ... has difficulties in bearing children. ... Thus you can recognize in the soul, the body and the fate, the same character; and this is indeed analogous to the constellations in such a way that it is impossible for the soul to be the molder of its entire fate, because fate is something coming from the outside, something foreign.[17]

The position of the Sun and the planets at his own birth had, he claimed, also affected his character:

> With me, Saturn and the Sun operate together. ... Therefore my body is dry and knotty, not tall. The soul is faint-hearted, it hides itself in literary nooks; it is distrustful, frightened, seeks its way through tough brambles and is entangled on them. Its moral habits are analogous. To gnaw bones, eat dry bread, taste bitter and spiced things is a delight to me; to walk over rugged paths, uphill, through thickets is a feast and a pleasure to me. I know no other way of seasoning my life than science; I do not long for other spices and reject them if offered to me.[18]

Kepler's boundless enthusiasm for (and insatiable curiosity about) every subject in which he took an interest resulted in his publishing a short book on astrology early in 1602, with the title "On Giving Astrology Sounder Foundations" (translated into English by J. V. Field[19]). As Field and others have noted, though, Kepler's Copernican stance makes his belief in astrology even harder to understand. It was one thing to believe that the planets influenced events on Earth when Earth was believed to be a special creation at the center of the Universe. It was quite another thing to carry on believing this when the Earth had been demoted to the status of just another planet. Why should the other planets have these influences when the Earth was no longer any different in kind from them? It is probably no coincidence that the gradual acceptance of Copernicanism in the decades after Kepler's death coincided with a general decline in belief in astrology.

In 1608, he was asked to cast a horoscope for an anonymous individual, who Kepler had correctly guessed was Albrecht von Wallenstein, an up-and-coming young soldier in Rudolph's army. As a result he was able to provide a horoscope for Wallenstein that matched well with Wallenstein's life. Many years later, this would stand him in good stead.

Even toward the end of his life, his belief in astrology continued. In his preface to the *Rudolphine Tables*, published in 1627, he said, "Although one may deny that *events* of human affairs depend on the stars, nevertheless he is certainly forced to recognize some *effects* on human affairs."[20]

KEPLER'S SUPERNOVA

The people of the late sixteenth and early seventeenth centuries were fortunate in one astronomical respect. Within the span of a single generation, they saw two supernovae in the night sky, one in 1572 and one in 1604. Supernovae are extremely bright, and apparently new, stars that appear suddenly and dramatically in the night sky (and can even be bright enough to be seen in the day), before slowly fading away over a period of a few months.

Notwithstanding the name (*nova* is the Latin word for *new*), supernovae are usually the catastrophic explosions that mark the end of a massive star's life. Any star that has a mass more than about eight times the mass of our own Sun will end its life in this way. The explosion is so violent that a typical supernova can, for a few weeks, shine as brightly as a whole galaxy of stars. Supernovae occur in our Galaxy at the rate of roughly two or three every century, but the vast majority of these are obscured from our view by the huge clouds of gas and dust that permeate the plane of the Galaxy. (We know the rate at which to expect supernovae in our own Galaxy by observing the rate in other galaxies.) We have been unlucky in not having been able to see any in our own Galaxy since 1604, and we are well overdue for another.

Supernovae are also an essential part of the process that eventually leads to life. The early Universe, some 13.8 billion years ago, contained only hydrogen and helium, the two simplest elements (plus negligible amounts of lithium, the next element up). Most of the remaining elements, required to make rocky planets like our own, and to make astronomers and other forms of life, came into existence only as a result of nuclear reactions in the centers of massive early stars. It takes the explosion of a supernova both to create even more elements and to release all these elements into interstellar space. Over many millions of years, these then slowly diffuse throughout the Galaxy, so that they eventually form a small part of the gas clouds that collapse to create later generations of stars— together with their retinues of rocky planets. Apart from hydrogen, all the atoms inside your body were created inside a star. We really are made of stardust—or, more prosaically, we are made of nuclear waste.

Although most supernovae are caused by the catastrophic gravitational collapse of a single old and massive star, a minority occur in binary star systems, in which two stars orbit around their mutual center of gravity. One of these must be a white dwarf, a smaller-mass star that has already reached the end of its life and is slowly fading away. However, its gravitational pull can still gradually suck in matter from its more diffuse companion star, and when it eventually reaches a certain critical mass, it, too, explodes as a supernova. (For rather obscure reasons, astronomers label these as Type 1a supernovae.)

Type 1a supernovae have the useful property that they all have the same intrinsic brightness. So by measuring their apparent brightness, we can calculate their distance. In 1998, these supernovae were used to measure distances to a large number of very remote galaxies. The result led to the astonishing conclusion that our Universe is not merely expanding (which had been known for the previous seventy years), but that the rate of expansion is actually accelerating, rather than slowing down (as had previously been thought). The three cosmologists who led the two teams that made this remarkable discovery were awarded the 2011 Nobel Prize in Physics for their achievement.[21]

None of this was known, or even suspected, by the astronomers of the sixteenth and seventeenth centuries. Any change in the night sky had been believed to be something that happened below the orbit of the Moon, since (in accordance with Aristotelian doctrine) everything from the Moon's orbit upward was meant to be perfect and unchanging. Tycho Brahe blew a colossal hole in this belief by demonstrating—on parallax grounds—that the supernova of 1572 must have been well beyond the Moon's orbit.[22]

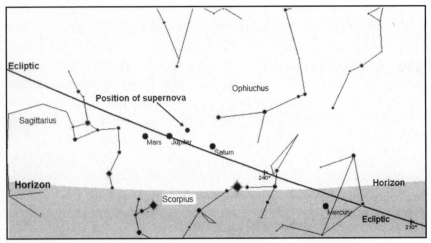

Figure 5.3. The location of Kepler's Supernova, very close to Mars, Jupiter, and Saturn, low on the western horizon in Prague, early in the evening of October 17, 1604 (SkyMap Pro software).

The supernova of 1604 is now known as Kepler's Supernova, because he observed it extensively and wrote about it. (*De Stella Nova—On the New Star*—dedicated, of course, to his patron and paymaster, Rudolph.) We now know that it was a Type 1a supernova. It was first observed in Prague by one Johann Brunowsky on October 10. The following morning, he informed Kepler, who initially doubted the report. Cloudy weather prevented him from confirming the account for a full week, but on October 17 the sky was clear and Kepler saw the new star for himself, in the obscure constellation of Ophiuchus. By complete coincidence, the planets Mars, Jupiter, and Saturn were all remarkably close by, strung together along the ecliptic (the Sun's annual path against the stars) almost in a straight line. Even Mercury was not far away, although it was below the horizon. To the superstitious people of the time, the appearance of a new star, and one so close to three planets, surely signified that something of supreme importance was about to happen—perhaps the overthrow of the Turks, the downfall of Islam, or even the Day of Judgment?[23]

GREAT CONJUNCTIONS

Even without the arrival of the new star, the appearance of the three outer planets (and Jupiter and Saturn in particular) so close to each other was in itself believed to be special. To an astrologer (although certainly not to a modern astronomer), there was a huge significance in the positions of the planets in relation to the background of the constellations of the Zodiac, all twelve of which lie along the ecliptic. For reasons best known to themselves, astrologers arbitrarily divided these twelve constellations into four "trigons": airy, watery, earthly, and fiery—named after Aristotle's four elements—and each containing three constellations. The fiery trigon was made up of the constellations of Aries, Leo and Sagittarius, which are separated from each other in the sky by some 120°, and was believed to have particular importance.

It was the general topic of Great Conjunctions (when Jupiter and Saturn are close to each other), which occur about every twenty years, that Kepler had been explaining to his pupils in Graz when he had the inspiration for his book *Mysterium Cosmographicum*. Each time Jupiter and Saturn approach each other, they are a little less than 120° away from their previous encounter.

So, as figure 5.4 shows, successive Great Conjunctions will occur at point 1, then point 2, and so on, jumping from point to point around the sky (relative to the background stars). About every 60 years, the Great Conjunction will be a few degrees away from where it had previously been (e.g., positions 1, 4, and 7 in figure 5.4) And about 800 years pass before a Great Conjunction again occurs in roughly the same part of the sky. Astrologers thought that the 800-year interval between occasions when a Great Conjunction again entered a member of the fiery trigon was of supreme significance and tried to tie it to major events

on Earth. In reality, of course, it is no more than an accidental consequence of the random geometry of the Solar System.

Kepler wrote about this 800-year interval in his *De Stella Nova*.
It was an integral—but totally mistaken—part of his religious beliefs
(and that of all his contemporaries) that the Earth and the Universe
were less than 6,000 years old. The creation, it was believed, had
occurred in roughly 4,000 BCE. The biblical evidence for this was considerable. In the New Testament, Luke, chapter 3, for example, sets out
a genealogical table of only seventy-six generations stretching back
from Jesus to Adam, believed to be the first man. Genesis, chapter 1
in turn states that Adam had been made on the fifth day of creation.
It seemed obvious from these passages, and others, that the Universe
could only be a few thousand years old.

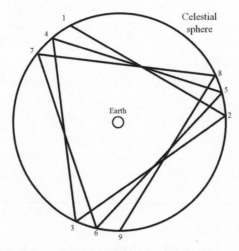

Figure 5.4. Successive Great Conjunctions occur about every 20 years
at point 1, then point 2, then point 3, and so on. Every 60 years, a
Great Conjunction will occur close to the one of 60 years previously.
And about every 800 years, a Great Conjunction will occur as Jupiter
and Saturn again enter one of the so-called fiery trigon zodiacal signs.

(In fact, anatomically modern *Homo sapiens* can be dated back
roughly 200,000 years, corresponding to some 8,000 generations,
rather than a mere 76. The Earth, in turn, is about 4.5 billion years old,
and the Universe is some 13.8 billion years old.)

So in *De Stella Nova*, Kepler endeavored to tie in the 800-year cycle to the time he believed marked the beginning of the Universe. As table 5.1 (translated from his book) shows,[24] while admitting that it was "not at all precise"[25] he tried to demonstrate that significant events had occurred every 800 years.

The reference to Rudolph in this table was no doubt intended to praise his patron. However, he also sent one copy of the book to King James I of England and included a flattering reference to James as the Philosopher King.[26] James was also referred to as Plato, and Kepler as Diogenes. It is difficult to see this as anything other than a first attempt to curry favor with a possible future patron if Rudolph II were to be replaced by somebody less agreeable.

Table 5.1. Kepler's attempt to demonstrate an 800-year cycle.

BC/AD (BCE/CE)	Significant people	Coincident events: reader, take care to note the effect of the trigons
4,000	Adam	Creation of the world.
3,200	Enoch	Robberies, cities, arts, absolute power.
2,400	Noah	The Flood.
1,600	Moses	Exit from Egypt. Law.
800	Isaiah	Era of Greeks, Babylonians, Romans.
	Jesus	Roman monarchy. Reform of the world.
800	Charlemagne	Empire of the West and the Saracens.
1,600	Rudolph II	Life, fates, & our destiny, of which we write here.
2,400		What will become of us and our flourishing Germany? And who will be our successors? Will they remember us? If however the world lasts this long.

THE DATE OF THE BIRTH OF JESUS

An illustration of Kepler's wide-ranging enthusiasm for anything that captured his interest can be seen in the huge effort he put into researching the date of the birth of Jesus. He published at length on the subject, first in 1606, then in 1613 (this time in German, for the benefit of the common people), and then again in 1614.

The division of our modern calendar into BCE (BC) and CE (AD) originated with the sixth-century monk Dionysius Exiguus. He argued that the year in which he produced an analysis of the dates of Easter was 525 years since the birth of Jesus, without making it clear how he had arrived at that figure. This assumed date for the birth of Jesus was in effect defined as being almost at the end of 1 BCE (on December 25), and all other historical events were later stated with reference to this assumed date. Since there is no "year zero," 1 BCE (BC)was immediately followed by 1 CE (AD).

The idea that Jesus was not after all born exactly at the turn of the century was first suggested in 1605 by Laurentius Suslyga, a Polish Jesuit academic who lived in Kepler's old hometown of Graz. In essence, the problem he had identified was that Herod the Great had died early in 4 BCE. ("Great" is something of a euphemism, as Herod's rule was thoroughly vicious and brutal.) Yet if the Gospel of Matthew is to be believed, Jesus was born while Herod was still alive. So Suslyga concluded that Jesus had probably been born around 4 BCE or earlier.[27] While on a visit to Graz, Kepler came across a copy of Suslyga's thesis and found himself "wondrously delighted" by his arguments, which he broadly accepted. He also saw implications in Suslyga's discovery that Suslyga himself had not spotted.

Two points in particular struck Kepler. First, he knew that in 7 BCE there had been another Great Conjunction. For most of 7 BCE Jupiter and Saturn had been very close to each other in the sky. By February of 6 BCE they had been joined by Mars, and together these three planets formed a bright triangle in the early evening sky. Kepler was deeply struck by the fact that, on Suslyga's new analysis of dates, this astrologically significant configuration must have occurred at most a year or two before the star that (according to the author of Matthew's Gospel, the only source for the story) had led the Magi to Bethlehem. Surely, he thought, God (who was clearly also a keen astrologer) must have arranged this coming together of these planets as a prelude to the appearance of the Star of Bethlehem.

Second, in December 1603, he had observed another of the 800-year conjunctions of Jupiter and Saturn, as they began a new cycle in one of the fiery trigons. Remarkably, a few months later, this, too, had been followed by a close approach of Mars and then by the appearance of Kepler's Supernova (see figure 5.3). The similarity of these events, some 1,600 years apart, was so great that he could not see them as a coincidence, and he drew a parallel[28] between the Star of Bethlehem and the new star of 1604.

There is an interesting modern twist in the story of the attempts to find the correct year of Jesus's birth. The birth stories are to be found only in the Gospels of Matthew and Luke, both of which were written around 80 CE. There are no references to them (not even passing references) anywhere else in the New Testament or in any non-Christian sources. Kepler and all his contemporaries automatically—and incorrectly—assumed that there was no contradiction between Matthew's account of the birth of Jesus and that of Luke. But Luke's Gospel has the birth of Jesus occurring when Quirinius became governor of Syria and at the time of a Roman census. We now know that both these events occurred in 6 CE (AD), and that there was no Roman census in this area before that.[29] So if the author of Luke is to be believed, the birth of Jesus was in 6 CE (AD), rather than 4, 5, or 6 BCE (BC).

KEPLER'S SOMNIUM (THE DREAM)

At various stages in his life, Kepler worked on what we would now think of as a science fiction story of what it would be like to journey to the Moon.[30] He started to write this as a student in his last full year at Tübingen,[31] in 1593, continued it while working on his *Optics* in 1604, and—to please his friend Johann Matthäus Wackher von Wackenfels (an adviser to the Emperor Rudolph)[32]—did further work on it in 1609. The story was not finally completed until 1630 and was not published until 1634, four years after his death, but the earlier versions have a chilling relevance to his mother's later trial for witchcraft.

The story itself is very short—only ten pages long in the original Latin—and, it has to be said, it is not particularly exciting. Its interest lies in the understanding Kepler demonstrates of what it must be like to live on other worlds, mainly in the 223 footnotes to the story. Kepler has a dream about an Icelandic man, Duracotus, who lived with his mother. His father had died at the age of 150, when Duracotus was three. When he was fourteen, his mother sold him to a sea captain, who sailed to Denmark to deliver a letter to Tycho Brahe. Duracotus stayed behind and remained with Tycho for many years, learning about astronomy. Eventually, homesickness caused him to return to Iceland. Here he was happily reunited with his mother, who now revealed that she was in touch with spirits who had the power to transport people to the Moon.

His mother summoned one of these spirits, who explained that travel between the Moon and Earth was only possible during the darkness of a lunar eclipse. In these conditions, travel was easy for the spirits but was still dangerous for humans. Overweight or unfit people could not be taken. (Kepler makes a point of saying that this therefore excluded most Germans.) The journey time was a mere four hours. Humans had to be anaesthetized so they did not feel the huge shock of being lifted off the Earth at high speed.

Kepler takes the opportunity of the story to get across several basic astronomical facts to his readers in a novel way. He makes use of the spirit to explain that one side of the Moon always faces toward the Earth, but that the Earth is forever hidden from the inhabitants of the far side of the Moon. Their daytimes and their nighttimes are each as long as fourteen Earth days of 24 hours. The inhabitants of the far side of the Moon (the Privolvans) experience total blackness during their nights. In contrast, the nights of the inhabitants of the near side of the Moon (the Subvolvans) are illuminated by the Earth, which goes through a series of phases (just as the Moon does to us). The Earth is roughly four times the diameter of the Moon, so it appears to the Subvolvans to be four times as big in the sky as the Moon does to us. Unlike the Moon, the Earth also rotates on its own axis every twenty-four hours, and the Subvolvans can also see this rotation. Temperature variations, especially for the Privolvans, are huge—their days are far hotter than the hottest parts of Earth, and their nights are far colder.

He uses the story as a means of disseminating new and perhaps strange concepts to his readers. Critically for his Copernican view, he records that the inhabitants of the Moon experience the Moon as stationary, just as the Earth seems stationary to us. This was a new idea, and he hoped it would persuade others that it was perfectly reasonable to accept the concept of a moving Earth, even though it seems to us to be stationary. As for the planets, their motions seemed far more complex from the Moon than from the Earth. Just as the apparent retrograde motion of the planets as seen from Earth is simply a mirror of the Earth's own motion, the apparent motion of the planets as seen from the Moon is further complicated by the motion of the Moon around the Earth.

The story is clearly to a large extent autobiographical—Duracotus's close relationship with his mother, his absent father, his meeting with Tycho Brahe, his homesickness, and his love of astronomy are all a reflection of Kepler's own life. The unfortunate part of the story lies in the fact that Duracotus's mother (for which we can read Kepler's mother) was able to contact spirits. This could be taken as an admission that Kepler's mother was a witch. This part of *Somnium* was later to cause poor Kepler acute embarrassment.[33]

THE SIX-CORNERED SNOWFLAKE

As the year 1610 drew to a close, Kepler was looking for a suitable New Year present for his friend Wackher von Wackenfels. With both his short sight and his intense curiosity, he had often noticed that, whenever snow fell, every single snowflake had six sides. There were no exceptions. Why was this the case? Why were there never snowflakes with, say, five or seven sides? Why were they always flat and never three-dimensional?

So he wrote his friend a short treatise on the mystery of why all snowflakes had six sides. The result was his little book *De Nive Sexangula* (*On the Six-Cornered Snowflake*), published in 1611. It makes play of the word *nothing*—the German for *nothing* is *nichts*, which is pronounced roughly the same as the Latin for *snow* (*nix*). The book doesn't solve the problem of why snowflakes all have six sides, but this was hardly surprising. (We now know that this is a consequence of the structure of the water molecule. But atomic theory—let alone the means to probe the shape of molecules—was totally unknown in

Kepler's day.) However, what it does do is to canter through a range of ideas and perhaps related subjects. For example, was there perhaps a connection with the hexagonal shapes in the honeycombs of beehives? These had clearly been constructed so as to optimize the use of space for storing honey in the hive.

His thoughts included reflections on the famous Fibonacci series:

1, 1, 2, 3, 5, 8, 13, 21, 34, 55, . . .

in which any number in the series is the sum of the two previous numbers, and which pops up in a number of places in nature.[34]

In science, it is often important to ask the right question. Kepler had managed to do this before, for example, in asking why planets moved more slowly the farther they were from the Sun. He had now done it again with this question about snowflakes. And even though he—inevitably—failed to get the right answer, it had caused many other interesting questions to enter his fertile mind.

THE GREAT COMET OF 1607: HALLEY'S COMET

Comets are dirty snowballs, mixtures of ice and dust. (The term "dirty snowball" was first coined by Fred Whipple in the 1950s.) They are deep-frozen time capsules that preserve a record of conditions during the formation of the Solar System about 4.5 billion years ago. Billions of them are thought to orbit the Sun in a gigantic spherical cloud (known as the Oort cloud, after the twentieth-century Dutch astronomer Jan Oort) far beyond the orbit of Pluto. Gravitational perturbations sometimes disturb a comet, causing it to change orbit and fall in toward the Sun. As it gets closer to the Sun, it gradually develops a tail of evaporating debris, which slowly disappears again as it returns to the region from which it came. The brighter and closer comets are visible from Earth.

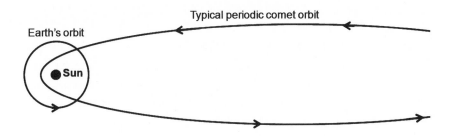

Figure 5.5. Elliptical (and hence periodic) cometary orbit.

One such comet appeared in the skies in 1607. Kepler saw it for the first time on September 26,[35] in the constellation of the Great Bear, and later wrote about it in his books *Bericht vom Kometen* and *De Cometis Libelli Tres*. As was often the case with his speculations, Kepler correctly made a good guess that there were many comets that went unobserved from the Earth,[36] although he realized that many people disagreed with him. What he never realized, however, was the power of his own laws of planetary motion. He knew his laws applied to the planets, but he did not know they applied to all the bodies in the Solar System (and beyond). Like many comets, the comet of 1607 actually orbits the Sun in a highly elliptical orbit, in accordance with Kepler's First Law, and sweeps out equal areas in equal times, in accordance with his Second Law. This means that it spends a long period of time traveling slowly through the furthermost reaches of the Solar System, before rapidly moving toward and around the Sun and then traveling back to the region from which it came.

Nearly 100 years later, the English astronomer Edmond Halley (1656–1742) used Kepler's laws (as later modified by Isaac Newton) to calculate that the orbits of the comets of 1531, 1607, and 1682 were very similar. He also noted that the appearances of these comets were about 76 years apart. He concluded that the three comets were one and the same object, and forecast that it would return 76 years after the 1682 appearance, in 1758. Halley died in 1742, but the comet duly reappeared on time and has ever since been known as Halley's Comet. Halley's prediction was a striking example of the increasing power of science to understand and predict the behavior of Nature.[37]

CORRESPONDENCE WITH DAVID FABRICIUS

During his time in Prague, Kepler corresponded with a large number of people. One of the most important of these relationships was with David Fabricius (1564–1617), a Lutheran minister and amateur astronomer who lived in the north of Germany. Fabricius acted as Kepler's sounding board when Kepler was writing his *Astronomia Nova*.[38] Today, his main claim to fame is for his discovery of the variable star now known as Mira. Mira (Latin for *wonderful*) is a red giant star in the obscure constellation of Cetus. Fabricius was the first person to notice that it varied regularly in brightness over a period of a little less than a year. It is visible to the naked eye for some four months, before it fades away and becomes invisible—except through a telescope—for about seven months. It was the first such variable star to be discovered and was another nail in the coffin of the belief that the heavens were unchanging. Fabricius himself came to an unfortunate end when he was murdered by an irate parishioner in 1617.

GIORDANO BRUNO

One significant figure of the time with whom Kepler did not correspond was Giordano Bruno (1548–1600). Bruno deserves a mention because his ideas were even more radical than Kepler's, and because those ideas have turned out to be essentially correct. He was a Dominican friar who strongly supported Copernicanism, but he went a lot further than other Copernicans of the age. He considered that the stars were suns, just like our own, and that they, too, had planets orbiting around them, and that these planets were also the home of intelligent life.[39]

We now know that Bruno was right on the first two counts, and we may one day know whether and to what extent there is intelligent life elsewhere in the Universe. The jury is still out on this, but given the unimaginably massive size of the observable Universe, it seems virtually certain that we are not unique. Bruno and Kepler probably never met, but Kepler often referred to Bruno in his writings, and he strongly disapproved of Bruno's idea of an infinite Universe in which the stars were suns like our own.[40]

Bruno was burned at the stake by the Roman Catholic Church on February 17, 1600. His execution was probably for theological reasons, but it is difficult to believe that his astronomical views played no part at all in the decision to have him killed. There is now a memorial statue of him on the spot where the execution took place, in the Campo de' Fiori—the Field of Flowers—in the center of Rome.

Kepler provides a little background on Bruno in a letter to his friend Johannes Brengger when he tells Brengger that "Bruno has been burnt in Rome; he is said to have been unyielding during the execution. He maintained the futility of all forms of religion."[41]

RUDOLPH

The Emperor Rudolph (1552–1612) was well satisfied with his Imperial Mathematician. In 1610, he declared "that we have graciously viewed, observed and reflected upon the faithful, diligent, proper and untiring most humble service which our mathematician, the faithful esteemed Johannes Kepler, now in his tenth year here, has with special pains most obediently exhibited and shown and daily performed to our agreeable and most gracious pleasure and satisfaction."[42]

Rather like many modern-day politicians in their attitude to science, Rudolph valued Kepler much as a drunk values a lamppost: more for support than illumination. (In his case, this was often in terms of Kepler's advice, dressed up as astrological prediction.) Nevertheless, in Kepler's entire career, Rudolph was probably his greatest admirer outside the scientific community.

CHAPTER 6

1610—
THE YEAR OF THE TELESCOPE

The arrival of the telescope fundamentally changed humanity's understanding of the Universe. For the first time in human history, people could move beyond the limitations of the unaided human eye to observe objects in the sky never seen before. Primitive by modern standards, the telescopes of the time must nevertheless have seemed as astonishing as the Large Hadron Collider seems to us today. It made what was seen in the sky far more real and no longer merely a matter for argument among academics. Following its invention in the Netherlands, probably by Hans Lippershey, in 1608, knowledge of the new device quickly spread throughout Europe.[1]

The first person ever to use a telescope to make systematic observations of the night sky was the wealthy English gentleman Thomas Harriot,[2] who used it on the evening of July 26, 1609, to make a crude drawing of the Moon. Harriot went on to make much better drawings of the Moon and to carry out a number of other astronomical observations. Unfortunately, he never published his results, and his contribution to astronomy was not appreciated until long after his death. There were good reasons for his reticence. Both of his patrons, Sir Walter Raleigh and Henry Percy, the ninth Earl of Northumberland, were (for different reasons) locked up in the Tower of London, and Harriot had no desire to call attention to himself and perhaps follow in their footsteps. He died, still largely unknown, in 1621, one of the first victims of the newly discovered tobacco plant. There is a plaque dedicated to him in the lobby area of the Bank of England on Threadneedle Street, close to where his grave (destroyed in the Great Fire of London of 1666) had been.

Harriot was more than rich enough to purchase a telescope. The as yet relatively unknown professor of mathematics at Padua University, Galileo Galilei, had to make his own. But Galileo, as well as being a brilliant theoretician and experimenter, was a skilled craftsman, and he produced a telescope (which he referred to as a spyglass) that was superior both in quality and in magnifying power to anything else in existence at the time. In November 1609, he turned a telescope that magnified twenty times on to the night sky and changed the nature of astronomy forever.

His early discoveries were recorded in his book *The Messenger from the Stars*, which he published early in 1610. Kepler's books were always lengthy, ponderous, and hard going. The many gems within them can be hard to find. In contrast, Galileo's book was short and punchy. It became an immediate bestseller. All 550 copies of the book were sold out within the first week, and Galileo turned from being a relatively obscure professor of mathematics into a noted international celebrity of his day. His discoveries shattered several deeply held preconceptions about the Universe.

His observations of the Moon showed "the surface of the Moon to be not smooth, even, and perfectly spherical, as the great crowd of philosophers have believed about this and other heavenly bodies, but on the contrary to be uneven, rough, and crowded with depressions and bulges. And it is like the face of the Earth itself, which is marked here and there with chains of mountains and depths of valleys."[3]

This was a blow to the Aristotelians, who had held that the Moon and the other heavenly bodies were perfect, unchanging, and fundamentally different from the Earth. By observing the lengths of shadows cast by lunar mountains and knowing roughly what the distance was to the Moon, Galileo was even able to use simple geometry to calculate the heights of these mountains. The Moon was not, after all, so very different from the Earth.

The ancient Greek Democritus (460 BCE—370 BCE) had been the first person to suggest that the Milky Way, the narrow band of light that can be seen stretching across the night sky (in modern times, alas, only in places far from sources of light pollution), might be composed of huge numbers of individual stars.[4] Galileo was the first person to show that this was the case: "For the [Milky Way] is nothing else than a mass of innumerable stars distributed in clusters. To whatever region of it you direct your spyglass, an immense number of stars immediately offer themselves to view."[5]

He also noticed that when he looked at the planets, they appeared as small discs, whereas when he looked at stars, they appeared essentially as unmagnified point sources and not as discs: "The planets present entirely smooth and exactly circular globes that appear as little moons, entirely covered with light."[6] "It is worthy of notice that when they are observed by means of the spyglass, stars . . . are seen not to be magnified in size in the same proportion in which other objects, and also the Moon herself, are increased."[7]

This provided useful support for the Copernican view. An argument against Copernicanism (deployed by Tycho Brahe, among others) had been that the stars ought to show some annual parallax if the Earth really did shift its position relative to the Sun and stars. Only if the stars were inconceivably distant would we not see this effect. The fact that Galileo's telescope did not magnify the stars was the first observational indication that they are indeed inconceivably distant.

But perhaps the most explosive discovery was of four small moons orbiting around Jupiter. "Four planets never seen from the beginning of the world right up to our day."[8] Until then, it had been possible to pour scorn on Copernicanism by asking why it was that, if the Earth was indeed just another planet, it was the only one to have a moon. Surely (it was argued) the Moon made us special—but not after Galileo's discovery. Moreover, the idea that one body (the Moon) could orbit another body (the Earth), which was in turn orbiting a third body (the Sun) was also thought not to be credible. Yet whether Jupiter orbited around the Earth or around the Sun, Jupiter's satellites were an exact parallel to the Moon, and demonstrated that this sort of motion was indeed perfectly possible. And finally, until this discovery, it had certainly seemed to the unaided eye as if everything in the sky was orbiting around the Earth. The moons provided the first example of objects that were demonstrably not orbiting around the Earth.

Galileo was anxious to move away from Padua and back to his native Tuscany. To do this, he needed a patron—Cosimo de' Medici, the Grand Duke of Tuscany. So he decided to call these new moons the Medicean Stars. The ploy worked, and within six months Galileo found himself back in Florence, working at the Grand Duke's court. But the names he had chosen didn't stick. We now call the Galilean moons by the names of Io, Europa, Ganymede, and Callisto, and they are four of the most fascinating bodies in the Solar System. They are by far the largest of Jupiter's many moons, which at the last count numbered over sixty.

The introduction to *The Messenger from the Stars* picked up on the fact that his observations favored the Copernican view of the Universe in a number of respects. In talking about the satellites of Jupiter, he finally nailed his colors to the Copernican mast by declaring that "all [the moons of Jupiter] together in mutual harmony complete their great revolutions every twelve years about the center of the world, that is, about the Sun itself."[9]

KEPLER'S RESPONSE TO GALILEO'S BOOK

Kepler first heard news of Galileo's discovery of Jupiter's moons in March 1610, when his friend Wackher von Wackenfels called on him to tell him. Kepler was at his home at 4 Karlova, in the center of Prague, when Wackher drew up in his carriage and gave him the news. Kepler's response gives a measure of the excitement he and many others must have felt over these wondrous new discoveries. "So great was my surprise at this piece of news ... as he with joy, I with shame, both with laughter because of the novelty of the situation, that I was hardly able to listen to the narrative."[10]

The report Wackher passed on was unclear on details. So Kepler's immediate concern was that Galileo had perhaps discovered four new planets going around the Sun. This would, if true, drive a coach and horses through the central thesis in his book of thirteen years earlier. *Mysterium Cosmographicum* had stated that there were only six planets because there were only five perfect solids. So Kepler suggested to Wackher that perhaps Galileo had seen four moons, one each around the planets Venus, Mars, Jupiter, and Saturn. Wackher had another idea: perhaps instead these four planets had been observed going around some of the fixed stars, just as Giordano Bruno and others had suggested.[11] In the event, they were both wrong, although Kepler's idea turned out to be closer to reality.

They didn't have to wait long to find out the truth. The Emperor Rudolph, who was keen to hear his opinion, allowed Kepler to have a quick look at his own copy. Then in April, Galileo, who was also anxious to hear Kepler's opinion on his discoveries, sent Kepler a copy of his book via the Tuscan ambassador in Prague, asking for his reaction. Many others, such as the Italian astronomer Giovanni Magini, had suggested that Galileo's results were an illusion, somehow created by the telescope. Not Kepler. Not only did he write back to Galileo immediately, accepting without question Galileo's results, he then published his letter in a little book, *Conversation with Galileo's Messenger from the Stars*. Contrary to what some suggested, there was not a word of criticism of Galileo in the book. Quite the converse—Kepler was characteristically generous in his praise for Galileo and his discoveries, differing only on minor points of detail (such as, for example, the reason for the Moon's reddish color during a lunar eclipse). He agreed that he would perhaps be seen as reckless in backing Galileo without direct experience of his own, but he found Galileo's whole style somehow convincing.[12] There was indeed something about the deliberately careful and detailed style of Galileo's book that made it extremely plausible.

The discovery of the four moons of Jupiter led Kepler to suggest (on symmetry grounds) that there must be two moons going around Mars, as well as six or eight going around Saturn, with perhaps one moon each for Venus and Mercury. The moons of the planets would then form either an arithmetic progression of sorts (1-1-1-2-4-6) or a geometric progression (1-1-1-2-4-8).[13] In reality, the number of moons going around each planet is entirely the result of random processes. This was another example of Kepler's erroneous use of numerology. But quite by accident, he turned out to be right about the planet Mars, which does have two tiny moons, Phobos and Deimos. These are far too small to have been detected with the primitive telescopes available in 1610 and were not found until 1877. They are now thought to be asteroids that have been captured by Mars.[14]

Kepler's fertile mind came up with several other comments on Galileo's findings. He suggested, for example, that it would be possible to measure the parallax of comets to an unprecedented level of accuracy by comparing their positions with the numerous faint stars that

could only be seen through a telescope.[15] This was perhaps unfortunate, because Galileo was never able to accept that comets are the very
distant bodies we now know them to be.

Kepler also suggested, on totally spurious grounds, that the Moon
was not a very dense body.[16] This was yet another occasion when his
speculations, based on entirely erroneous reasoning, nevertheless
turned out to be correct—we now know that the Moon is a lot less dense
than the Earth. We think this is because the Moon originally formed from
debris that came from the Earth's crust (which is lighter than Earth's
interior), when the Earth was hit by a planet the size of Mars during the
chaotic formation of the Solar System some 4.5 billion years ago.[17]

He had originally thought of the bright areas on the Moon as seas
(and had argued as much in his *Optics*, published seven years earlier[18]),
so he differed from Galileo and others, who thought instead that the
dark areas were seas. But Galileo's arguments persuaded him to change
his mind. In fact, virtually all parts of the Moon's surface are bone-dry.
(The exceptions are the bottoms of lunar craters at the poles; these are
always very cold because they never receive any sunlight. We now know
they contain some frozen water, deposited there by comets over the last
four billion years.) But the word has stuck; astronomers still refer to the
dark areas by the Latin word for *seas—maria*. He also speculated about
the possibility of life on the Moon, mentioning that he had written about
this on previous occasions.[19] He accepted the conclusion of Galileo and
others that there was air, and even rain, on the Moon,[20] although we now
know this is not so. And it seemed virtually certain to him that there was
life on Jupiter too, on the grounds that there had to be people (albeit
inferior to us) to appreciate the four moons of Jupiter, otherwise God
would not have put them there.[21]

It is now clear that neither the Moon nor Jupiter is a host for intelligent life. Of all the numerous sites in our Solar System, only Earth
has had the right environment for intelligent life to evolve. However,
it is at least a serious possibility that there is primitive bacterial life
below the surface of Mars, in the ocean we now know exists below the
icy surface of Jupiter's moon Europa, and perhaps also below the icy
surface of Saturn's moon Enceladus.

Kepler's friend Wackher cleverly pointed out another feature of

Jupiter—that it must rotate.[22] Kepler had, to his own satisfaction, demonstrated in *Astronomia Nova* that it was the Sun's rotation that caused the planets to revolve around it. (In fact, the Sun's rotation has nothing to do with it.) On the same (faulty) logic, it therefore followed that as Jupiter had satellites revolving around it, then it, too, must rotate. In fact, Jupiter does rotate very rapidly (roughly once every ten hours), but this again has nothing to do with keeping its satellites in orbit.

Kepler recognized that the stars shone by their own light, in contrast to the planets, which depended on reflected light. But in spite of Galileo's estimate that there must be some ten thousand stars visible through his telescope (with the implication that many more could be seen through an even more powerful telescope), Kepler strongly rejected Giordano Bruno's idea of an infinite Universe, in which the stars have the same nature as our Sun and are also orbited by planets. He regarded this proposal as "horrible"[23] and disagreed violently with Wackher, who was a supporter of the idea. He believed in a finite Universe and put forward what was (in pre-telescopic times) a very clever argument in favor of this belief. The stars—it was wrongly thought—all appeared to have a small diameter as seen from Earth. Kepler argued that if these surfaces were added up, they would collectively create a bigger disc than the Sun, yet it is clear that collectively they are much less bright than the Sun.[24] (Otherwise the night sky would be as bright as the sky in daytime.) It followed that the individual stars must be considerably less bright than our Sun, which was therefore special. If the stars were farther away, this would make their true diameters even larger, but their light output relative to their size would be smaller and they would be even less like our Sun. The Solar System therefore remained, for Kepler "the principal bosom of the Universe."[25] The argument falls down because Kepler failed to take into account the implication of Galileo's telescopic discovery that the stars appeared as point sources, with no apparent diameter. In reality, many stars are just as bright as, or considerably brighter than, our Sun.

His argument was closely related to—and was probably the stimulus for—what is now known as Olbers' paradox: Why is the sky dark at night? The paradox is as follows: the stars in our Galaxy (and beyond them the billions of other galaxies) are randomly distributed in space, but they can be thought of as occupying a series of concentric spheres around the Earth (figure 6.1). Each (imaginary) sphere will emit light in proportion to the surface area of the sphere, which is in turn directly proportional to the square of the sphere's radius. On the other hand, light received by the Earth from any particular sphere is inversely proportional to the square of the sphere's radius (as first demonstrated by Kepler).[26]

The result of this is that the greater amount of light emitted by an outer sphere (compared with an inner one) is exactly canceled out by the greater distance from Earth. The consequence: the light reaching Earth from any one sphere will be exactly the same as the light reaching Earth from any other sphere, regardless of distance. More and more spheres will cause more and more light to reach Earth.

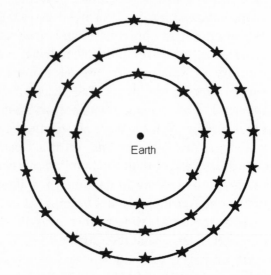

Earth

Figure 6.1. Olbers' paradox—stars and galaxies beyond them can be imagined as a series of concentric spheres around the Earth.

Olbers' paradox (named after the nineteenth-century German astronomer Heinrich Olbers) states that if the Universe has existed forever, is infinitely large, and is uniformly filled with stars (and galaxies beyond them), then the light we receive from the stars and galaxies should be infinitely great.

But it isn't. The one obvious fact about the night sky is that it is dark. Therefore either the Universe as we see it is finite in extent or it hasn't existed forever (or both). It is a remarkable fact that such a profound conclusion can be deduced from such a simple observation.

Nowadays, we know that the resolution of the paradox lies in the twin facts that our Universe has a finite age and that it is expanding. The finite age means that, even if the Universe is infinite in extent, the light from the outer spheres will not yet have had time to reach Earth. In addition, the fact that the Universe is expanding causes light from more distant sources to be red-shifted. This weakens the light, and in effect means that we receive less and less light from outer spheres. There is a horizon beyond which the Universe is expanding so fast that light from the outermost shells can never reach us.

But Kepler had clearly been rattled by Bruno's suggestion that the stars are just suns like our own, with planets like our own, and are inhabited by other intelligent beings like ourselves. Suppose Bruno were right? How else could the superiority of our position in the Universe be demonstrated? The answer lay, he believed, in his five perfect solids. It was his perfect solids that made the Solar System unique. God could not possibly have used the same pattern twice, and any other pattern was inferior. So even if these other planetary systems existed, they were demonstrably inferior to our own.[27]

In one section of the book, in a wonderfully prophetic sentence in which he seems to foresee the eventual coming of manned space travel, Kepler writes: "Given ships or sails adapted to the heavenly breezes, there will be those who do not fear even that vastness."[28]

Galileo was deeply grateful to Kepler for his support, and he wrote—a little belatedly—in August 1610 to tell him so. He also took the opportunity to describe the difficulties he had in persuading people even to look through his telescope:

> What is to be done now? Shall we follow Democritus or Heraclitus? We will laugh at the extraordinary stupidity of the crowd, my Kepler. What do you say to the main philosophers of our school who, with the stubbornness of vipers, never wanted to see the planets, the Moon or the telescope, although I offered a thousand times to show them the planets and the Moon. Really, as some have shut their ears, these have shut their eyes towards the light of truth. This is an awful thing, but it does not astonish me. This sort of person thinks that philosophy is

a book like the Aeneid or Odyssey, and that one has not to search for truth in the world of nature, but in the comparisons of texts.[29]

After the publication of *The Messenger from the Stars*, Galileo went on to make other discoveries. In July 1610, he found that the planet Saturn did not always appear as a circular disc. Instead, it seemed to have two handles, one attached to either side. His telescopes were never good enough to resolve these handles into the ring system that we now know orbits Saturn. This was a discovery that would have to wait for Dutch scientist and mathematician Christiaan Huygens, forty-five years later, by which time the quality and magnifying power of telescopes had significantly improved. Galileo had a cunning way of announcing his discovery. In the days before scientific journals, there was nothing to stop somebody else from claiming he had already discovered Saturn's strange shape if Galileo simply announced what he had found. So he circulated his finding as an anagram:

Smaismrmilmepoetaleumibunenugttauiras[30]

the meaning of which would be revealed at a much later date, making subsequent priority claims from others much less plausible. The anagram (with u's and v's interchangeable) can be rearranged to

Altissimum planetam tergeminum observavi

which translates as

I have observed the highest planet [i.e., Saturn] *in triplet form.*

When Kepler got hold of the anagram, his inventive mind rearranged it as

Salve umbistineum geminatum Martia proles.

which translates as

Hail flaming twins, offspring of Mars.

So until he found out the correct rearrangement, Kepler concluded—not unreasonably—that Galileo had discovered the two satellites of Mars that Kepler had already suggested Mars must have on symmetry grounds.

Galileo also found that, over a period of several months, Venus showed phases, rather like the Moon does. At first sight, this may not appear particularly significant. But in fact it provided the first observational test of a difference between the Ptolemaic (Earth-centered) system and the Copernican (Sun-centered) system. The point is illustrated in figure 6.2. If the Ptolemaic system is correct, then Venus orbits around the Earth, while also going around its epicycle. It also always stays closer to the Earth than the Sun does. This should lead to the Ptolemaic set of phases shown in figure 6.2.

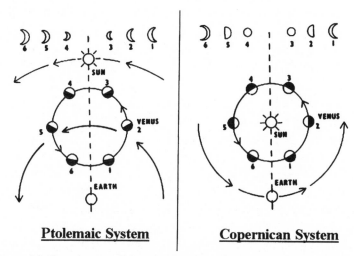

Ptolemaic System   **Copernican System**

Figure 6.2. The phases of Venus provide an observational test of the difference between the Ptolemaic and Copernican systems.[31]

In contrast, if the Copernican system is correct, then sometimes Venus is somewhere between the Earth and the Sun, and sometimes Venus is somewhere on the far side of the Sun. Crucially, this leads to a different set of phases (figure 6.2 again). Galileo found that he saw the phases expected in the Copernican system. So he had demonstrated that at least this aspect of Copernicanism was true and at least this aspect of the Ptolemaic system was false.[32] (The compromise system devised by Tycho would also have survived this observational test, because according to Tycho, Venus and the other planets orbit the Sun, which in turn orbits the Earth. The Copernican and Tychonic systems were geometrically, if not physically, equivalent.)

Once again, Galileo announced this discovery in the form of an anagram, which could be rearranged to form

Cynthiae figures aemulatur mater amorum.[33]

which translates as

The mother of love [Venus] *emulates the shapes of Cynthia* [the Moon].

Kepler had several goes at rearranging the anagram. One of these was

Macula rufa in Jove est, gyrator, etc.[34]

which translates as

There is a rotating red spot in Jupiter.

This was an astonishingly lucky guess, because of course Jupiter is now famous for its Great Red Spot. The visible surface of Jupiter isn't solid—what we see is the top layer of its extensive atmosphere. The Great Red Spot is a prominent atmospheric feature that has survived for at least the last 180 years and does rotate in a counterclockwise direction roughly every seven days. But there was no way that Kepler could have known of its existence, let alone the fact that it rotates. It was first definitively discovered by German astronomer Heinrich Schwabe in 1831, but it may have been seen a lot earlier by the Italian astronomer Jean-Dominique Cassini in 1665 (with a much better telescope than was available in Galileo's time).[35]

Galileo also discovered sunspots on the surface of the Sun. This was one of those discoveries that was in the air at the time. Johannes Fabricius, the son of David Fabricius, discovered them independently sometime in 1611, as did Christopher Scheiner (who argued that he had in fact found them first). Thomas Harriot also discovered sunspots independently, late in 1610. Sunspots were another blow to the Aristotelian view of the Universe. The Sun was meant to be a perfect body. But how could it be perfect if it was covered with spots? This was probably one of the reasons why Scheiner and others wrongly argued that the spots were actually satellites of the Sun, rather than the surface features that we now know them to be.[36]

Unknown even to himself, Kepler had beaten them all to it. On May 28, 1607, he thought he had a chance of observing a predicted transit of Mercury across the face of the Sun. (Such transits occur roughly a dozen times in a century. The next is due in 2016.) He set up an apparatus that produced a projection of the Sun into a darkened room, and—lo and behold—saw a dark spot on the Sun's disk. He thought he was observing a transit.[37] Only when, a few years later, he heard about sunspots did he realize that he had actually observed one himself.

KEPLER'S REPORT ON HIS OBSERVATIONS OF THE FOUR SATELLITES OF JUPITER

In September 1610, Kepler published a report on his own observations of Jupiter's satellites. He had been loaned a telescope by Elector Ernst of Cologne, Duke of Bavaria, who was visiting Prague. It was in this report that he first introduced the term *satellite* to describe a moon going around a planet, and his terminology has been used by astronomers ever since. He was clearly aware of the important scientific principle of the need for independent observations. He and a group of friends all made independent notes of what they saw through the telescope before sharing these notes with each other.[38]

DIOPTRICE

Following directly on from the invention of the telescope, Kepler set himself the task of working out the laws governing the passage of light through lenses. He published the results in his *Dioptrice* in 1611. One of his most important findings related to the construction of a telescope.

The early telescopes were all refracting telescopes. In other words, they all used two lenses, an objective lens to gather the light and an eyepiece to focus it. Galileo's telescopes consisted of a convex objective lens and a concave eyepiece. This had the advantage of producing an image that was the right way up, but the disadvantage of giving a narrow field of view. The result was that the Galilean telescope was difficult to use—it was little short of miraculous that Galileo managed to get the results that he did from it.

Kepler improved considerably on the design by using two convex lenses. This had the disadvantage (not usually important for an astronomer) of producing an upside-down image (although this can be corrected by the insertion of a third lens), but a more than counterbalancing advantage of a wider field of view. Refracting telescopes ever since have been based on Kepler's design. If we are to give credit where it is due, we should refer to modern refracting telescopes as Keplerian telescopes.

FAME

Kepler was becoming increasingly famous throughout Europe. In 1611, his contemporary the English metaphysical poet John Donne anonymously wrote the work *Ignatius His Conclave*, a lengthy satire on the Jesuits. It included several references to Copernicus and the following passage on Galileo and Kepler:

> Of which, I thinke it an honester part as yet to be silent, than to do Galileo wrong by speaking of it, who of late hath summoned the other worlds, the Stars, to come neerer to him, and give him an account of themselves. Or to Keppler, who (as himselfe testifies of himselfe) ever since Tycho Braches death hath received it into his care, that no new thing should be done in [the skies] without his knowledge."[39]

CHAPTER 7

LINZ (1612–1626)—
THE HARMONY OF
THE UNIVERSE

W ith the publication of *Astronomia Nova*, 1609 had been the year of Kepler's greatest triumph. The year 1611 was to be his annus horribilis.

Political problems had been steadily mounting over the previous years. Rudolph was becoming ever more withdrawn and less able to cope with the increasing religious tensions in his territories. In 1608, he had been forced to cede control of Hungary, Moravia, and Upper and Lower Austria to his ambitious younger brother, Matthias. In 1609, a group of Protestants had forced their way into Rudolph's palace and managed to wrest further religious concessions from him—set out in the so-called Letter of Majesty. From 1611, Rudolph was held as a virtual prisoner in his palace by Matthias's army—his position as emperor had become untenable. And Kepler's salary as Imperial Mathematician fell further into arrears. It was hardly surprising that he started to look around for alternative employment.

But it was disasters in his family that were about to cause Kepler huge distress. His first two children had died in infancy many years previously, while he was still in Graz. Only his stepdaughter Regina had accompanied Kepler and his wife to Prague. On arriving in Prague in 1600, he and his wife had started a new family. During the critical years when he was working on *Astronomia Nova*, all went well. Susanna was born in 1602, Friedrich in 1604, and Ludwig in 1607. The children, although welcome, were a distraction from his work. He wrote to his friend Herwart von Hohenburg about the disruption caused by the birth of Friedrich:

The domestic disturbance which has produced this confusion (I must mention it in order to be fully excused) is caused by the women. What trouble, what fuss and what disturbance is created by inviting 15 to 16 women to my wife in childbed, to be hospitable and to see them to the door, etc. You must understand that on 3rd December a son was born to me and baptized the day before yesterday.[1]

Regina, then about eighteen, had married in 1608 and left the family home.[2] Then early in 1611, all three children contracted smallpox, just as their father had done at a similar age. Susanna and Ludwig recovered. The middle child, Friedrich, the favorite, did not. He died on February 10, at the age of six. His death had a devastating effect on both parents.

After Friedrich's death, Kepler became even more anxious to return to his native Württemberg. In March, he wrote to Johann Friedrich, the new young Duke of Württemberg, begging the duke to help him find employment in his fatherland as "a professor of philosophy or in a political position (in which I would have a little peace to complete my treasured philosophical studies)."[3] He was unable to resist pointing out that he had also received job offers from elsewhere—this may have been a reference to the possibility of a post for which Galileo had recommended him at Padua University (following Galileo's own departure for Florence), but this never came to anything. To make quite sure the duke got the message about his wish for a position, he also wrote to the duke's mother, Sybilla, politely asking her to persuade him. If this didn't work out, then perhaps Sybilla "would give me a possible position in your services (in which I should nevertheless have a little time for research and book writing)."[4]

The requests came to nothing. The duke would have been happy to provide Kepler with a post somewhere, but—as a result of Kepler's confession of religious doubts to the duke two years previously—the religious authorities in Württemberg were adamantly opposed to the idea. In a letter of April 25, 1611, they declared that Kepler's heretical views meant that he could not be seen as a "brother in Christ."[5] Kepler's final attempt to return to his native land had come to nothing. It was another overwhelming shock.

At last, in May 1611—just after Rudolph had been forced by Matthias to surrender control of Bohemia, and had in effect been relieved of the last of his powers—Kepler was offered the job of district mathematician in Linz, in Upper Austria, in a post that seems to have been created especially for him.[6] This pleased his wife, who had never liked Prague, and who in Linz would have been much closer to her own hometown of Graz. But a little over one month later, his wife died, probably as a result of an infection she picked up from Matthias's soldiers, who had been flooding into Prague. It was the third enormous blow to hit Kepler in the space of less than six months, and it totally shattered his normally optimistic view of life. He must have felt utterly overwhelmed.

A year later, Kepler wrote to a friend, Tobias Scultetus, who was an adviser at the court of the new emperor Matthias, about this tragic period in his life:

> Apart from the public misfortunes and the threats from outside, disaster in my own home has come over me in many ways. . . . I had a life companion, I do not want to call her my dearest one, for that is always or should always be the case; no, a woman to whom public opinion offered the palm of respectability, righteousness and modesty. In a rare way, she combined these virtues with outward beauty and cheerfulness of mind. Not to mention her inward virtues, her piety towards God, and charity towards the poor. With her I had blossoming children, especially a six year old boy, very much like his mother. . . . The boy was so tenderly united with his mother that one could not say that both were weak with love for each other, but rather mad with such love. I had to witness how, in her prime, my wife for three whole years was continuously afflicted with attacks of the raging tumors in her body, how she was shaken and in the end was so shattered that she not seldom was mentally deranged and quite out of her mind. And just when she seemed to be recovering, she was thrown back into a depression over one illness after another of her beloved children, and was wounded to the depths of her heart by the death of the little boy who was half her heart to her. Paralyzed by the atrocious deeds of the soldiers, and eyewitness of the battle in the town; driven to despair for a better future, and out of the inextinguishable longing for her lost darling, in the end she caught the Hungarian fever (her charity took revenge on her, as she would not

stop looking after the sick). In a melancholic depression, the saddest mental condition under the sun, she at last breathed out her soul.

Why do I speak of all this? Am I the only one who is treated cruelly by fate? Well, please learn from my report about my state of mind, since to their astonishment, some think I no longer show that elasticity which speculations in astronomy demand.[7]

Rudolph, although no longer emperor, valued Kepler's services and persuaded him to stay in Prague a little longer. Then he died in January 1612. Kepler was not the only one owed money by Rudolph. On his death, the former emperor was in debt to a total of some two and a half million florins to his former employees.[8] Matthias, who was elected as the new emperor, reconfirmed Kepler in the post of Imperial Mathematician, but with no requirement for him to stay in Prague.[9] So in April, Kepler left Prague for Linz. He deposited his two motherless children in the care of a widow he knew and traveled on to Linz alone. He was to stay here for nearly fifteen years, longer than anywhere else in his life.

Figure 7.1. Map of Upper Austria and surrounding area in 1600 (Euratlas maps).

Compared with Prague, Linz was something of a backwater. Kepler's job description—written especially for him—included teaching at a small school in the town, producing a map of Upper Austria, and carrying on with research into more or less whatever he wished.[10] In addition, he had the continuing burden of completing the *Rudolphine Tables*.

RELIGIOUS CLASHES

After such a horrendous year in 1611, Kepler was surely entitled to some time to get his life back together again. Once again, though, religion was to cause problems. As early as 1609, he had petitioned the Duke of Württemberg (whose permission he had to seek under the terms of his original agreement as a student at Tübingen University), asking to be allowed to seek employment outside Prague.[11] But Kepler, always honest and forthright in his pronouncements, had gone on to inform the duke that he was not able to subscribe to parts of the Formula of Concord. This document, published in 1577, had become the touchstone of Lutheran orthodoxy,[12] and one of its key points was a rejection of the Calvinist view of the Eucharist.

The modern mind has difficulty in understanding either the minutiae of religious doctrines or the intense hatred that small differences in these doctrines engendered. Nevertheless, they are a very important part of Kepler's story because the doctrine of the Eucharist caused him huge anguish throughout his life, so it requires some explanation.

The Roman Catholic belief was that during the sacrament of the Eucharist (also called Holy Communion), the substance of the bread and wine literally became the body and blood of Jesus, even though the physical appearance remained that of bread and wine. This is the doctrine of transubstantiation. Martin Luther believed that this doctrine was unscriptural and false. He nevertheless held that the bread and wine given at the Eucharist indicated the presence (in more than just a symbolic sense) of the body and blood of Jesus—the doctrine of the Real Presence (sometimes referred to as consubstantiation).

John Calvin and his allies dismissed both these doctrines, instead holding that the Eucharist was simply a form of memorial and thanks-giving service. The bread and wine were mere symbols. It was in part because of this difference in doctrine that many Lutherans regarded Calvinists as even more heretical than the Roman Catholics. Rather than criticizing God (for not having made the correct doctrine suffi-ciently clear in the Bible), all three sects were furious with the other two for taking what they saw as a heretical stand on this issue.

In his typically independent and thoughtful fashion, Kepler—although still a devout Lutheran—had concluded that the Calvinist doctrine made the most sense. His earlier honest but naive confes-sion to the duke of his doubts—which was bound to leak out—simply stored up trouble for him on his arrival in Linz.

Most people in Upper Austria were Lutherans, and the Lutheran Church there was run by one Daniel Hitzler, another former student of Tübingen University and some five years Kepler's junior. Hitzler was aware that Kepler's religious beliefs were not entirely orthodox, and he asked him to give his agreement in writing to the Formula of Concord before he could be allowed to take part in the Eucharist. Kepler explained that he was unable to do this. Hitzler was a hard-line Lutheran, so he refused Kepler's request for communion. Kepler appealed against his decision to the church authorities in Württem-berg, who sided firmly with Hitzler; Kepler's confession to the Duke of Württemberg a few years before had made this inevitable. The disagreement became common knowledge in Linz, and Kepler was branded by many as a wicked heretic. There were even times when his safety was at risk because of this.[13]

The struggle continued for many more years. In 1618, Kepler tried to get the support of his old tutor Matthias Hafenreffer, who was now the vice-chancellor of Tübingen University. He failed. Hafenreffer's final letter on the subject, written with the full agreement of the theo-logical faculty and only a few months before his death, closed the door forever to a reconciliation with the Lutheran Church. If Kepler did not change his mind on this matter, he was risking eternal damnation:

> Either you will cease your erroneous and entirely wrong chimera and embrace the sacred truth of humble belief, or else steer clear of the congregation of our church and our creed.[14]

Kepler's first few months in Linz were not all so ill-fated. It was at this time that he met Matthias Bernegger, an academic with a deep interest in astronomy, and some ten years younger than Kepler. Bernegger happened to be passing through the town on his way to Strasbourg to become the professor of history at the university. Although they never met again, they started a regular correspondence, and Bernegger was to become Kepler's best and most loyal friend.[15] (Over forty of Kepler's letters to Bernegger survive, as do over twenty of Bernegger's letters to Kepler.)

The following years in Linz were to be marked by the intertwining themes of family tragedy, further astronomical publications, and the chaos and destruction of the Thirty Years' War.

SECOND MARRIAGE

One aspect of his private life in Linz did eventually come right for Kepler. One of his first tasks on settling there was to find a second wife, if only to have someone to care for his two surviving children. The tortuous process by which he finally settled on a new bride was set out in a letter he wrote to a friend, Baron Strahlendorf in Prague, a week before his wedding.[16] The stark contrast between his deft approach to many scientific questions and his flat-footedness in trying to decide which of eleven possible candidates to choose from is one of the things that makes Kepler such an endearing character.

He had rejected the first and second candidates, only to have a third recommended to him by the wife of a friend. But Frau Helmhard's advice on the third candidate was sufficient to cause Kepler to decide instead to settle for the fourth candidate, while being annoyed that he had let the fifth slip away. Alas, the fourth had already grown tired of Kepler's constant hesitations and had instead accepted the proposal of another man, who had been emphatically asking for her hand for a long time. Kepler was now "as much annoyed about the loss of the fourth as I had been about losing the fifth." He went on to consider a further six candidates over the course of a year. Finally, however, he returned to the fifth, who had again become available.

Susanna Reuttinger, the fifth candidate and some eighteen years Kepler's junior, was a more suitable wife than Barbara Müller had ever been. She was an orphan from a poor background—her father had been a carpenter. But, unlike her predecessor, she had had some education, and Kepler considered that she had "a brain capable of learning what is still lacking." After his previous experience, he could see the advantage of an educated and intelligent life companion. The wedding was held in Susanna's home town of Eferding, a few kilometers west of Linz, on October 30, 1613. Afterward, there was a reception at an inn on one corner of Eferding's main square. (Four hundred years later, the building is still standing, and a plaque on the front wall proclaims to visitors that this was the site of their celebrations.) Their first child, Margareta, was born just over a year later, in January 1615.

Image 7.1. The building in Eferding where Kepler's wedding reception was held.

In marrying twice, Kepler was the exception among the four towering figures in astronomy of the sixteenth and seventeenth centuries. Copernicus never married (although there was much gossip about his relationship with his housekeeper). Galileo had a long-term mistress, who bore his three children. Newton also never married—possibly because he was such an unpleasant individual.

However, the marriage was still devastated by the appallingly high levels of child mortality common at the time. Margareta died in September 1617, at the age of only thirty-two months. Susanna was to produce a total of seven offspring, but five of these would die in infancy or childhood. Of Kepler's twelve children, only his daughter Susanna by his first marriage (born 1602), Ludwig, Cordula, and (possibly) Anna Maria survived to adulthood.[17] In an age when about two-thirds of all babies born would be dead before the age of twenty, this was just par for the course.

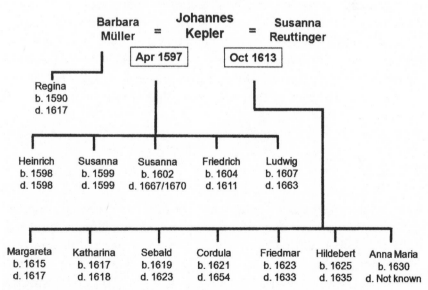

Figure 7.2. Of Kepler's twelve children, eight did not live beyond infancy or childhood.

NOVA STEREOMETRIA DOLIORUM VINARIORUM

One of the most important mathematical inventions ever is what is known as calculus. It enables the easy solution of otherwise intractable problems. One part of calculus (differentiation) allows the calculation of the gradient of a curve. The other (involving the opposite process: integration) allows the calculation of the area under a curve. The method was invented at about the same time (roughly forty years after Kepler's death) by both Isaac Newton and Gottfried Leibniz and led to a bitter dispute as to which of them had got there first.

In fact, there had been forerunners to the work of Newton and Leibniz. One of these was Kepler's first publication, in 1615, after arriving in Linz. It was a lengthy document titled *A New Means of Measuring the Volume of Wine Barrels*. In 1613, he was stocking up his home with wine from the exceptionally good harvest for that year. Ever curious, he became fascinated by the way wine sellers measured the volume of wine in a barrel. They did this simply by putting a measuring rod into a wine barrel at an angle and noting the maximum length of rod that would go into the barrel. Could this be justified? How could the measurement of a length possibly be a valid way of measuring a volume? Kepler conducted a very thorough mathematical investigation of the problem and demonstrated that—for these particular casks—the measuring rods did indeed measure the volume correctly. Some of the techniques he employed to reach this conclusion were precursors of the techniques that would become part of calculus.[18]

GALILEO

The initial opposition to Copernicanism had come largely from the Protestant churches. Now it was the turn of the Roman Catholics. Galileo had been a Copernican since 1597 or earlier, but he had kept his beliefs to himself. Only in 1610, with the evidence from his telescopic observations, had he felt able to emerge from the closet. His book *The Messenger from the Stars* made explicitly Copernican references and in subsequent years Galileo sided ever more strongly with the Copernican viewpoint. Complaints about his views became strong enough for the Congregation of the Holy Office (previously known as the Inquisition) to investigate the matter. On February 19, 1616, their advisers came to the following conclusion:

> [That] the Sun is the center of the world and completely immovable of local motion is foolish and absurd in philosophy, and formally heretical, inasmuch as it expressly contradicts the doctrine of the Holy Scripture in many passages, both in their literal meaning and according to the interpretation of the Fathers and Doctors.[19]

So Galileo was told that he could not hold or defend Copernican views any longer, and Copernicus's book *De Revolutionibus* was suspended "until it is corrected." As part of the general onslaught on heretical material, the first part of Kepler's *Epitome of Copernican Astronomy*, published in 1617, was also banned by the Church in 1619.[20] (As is always the case when something is banned, it simply encourages people to read the banned work. Vincenzo Bianci, an Italian who corresponded with Kepler, assured him that this would be the effect of the ban in Italy.[21])

Given the crackdown on his publications, it is probably just as well that Kepler did not accept the offer of the professorship of mathematics at Bologna University, following the death of the incumbent, Giovanni Magini, in 1617:

> I could only with the greatest difficulties transplant my living quarters from Germany to Italy. I feel tempted by the honor linked with this highly renowned place, the venerated professorship of the University of Bologna; and I am certainly attracted by the chance of getting a larger audience and a better position, and privately because of the financial advantages. Yet the period of life has passed when one feels stimulated by new circumstances and longs for the beauty of Italy or the promise of its long lasting enjoyment.[22]

HARMONICES MUNDI (THE HARMONY OF THE UNIVERSE) AND THE THIRD LAW

In the light of all the stresses in his life—the religious clashes, his tragic family life, and the outbreak of the Thirty Years' War, it is astonishing that Kepler was able to continue working at all. But the sheer misery of his life over these years drove him to try to find some escape, by constructing ever more fanciful notions of harmony in planetary motions. The year 1619 saw the publication in Linz of *Harmonices Mundi* (*The Harmony of the Universe*).[23] The work begins with an extravagant dedication to King James I of England. Kepler had previously sent a copy of his book on the supernova of 1604 to James, for whom he clearly had a great admiration. He was probably still toying with the idea of finding employment in England if life became too unbearable in continental Europe.

In 1620, he was invited to go to England by the English ambassador to the Habsburg court in Vienna, Sir Henry Wotton (famous for his quip that an ambassador is an honest man sent abroad to lie for his country). But he eventually declined.

> Mr. Wotton has shown great friendliness towards me; he regretted that he had to continue his journey so hurriedly. He told me to come to England. Yet I do not think I ought to leave this second home of mine, especially now, when it is suffering so much insult.[24]

Wotton in his turn was very impressed with Kepler, and with his *Harmonices Mundi*, and wrote to Francis Bacon about his meeting. (Bacon was a talented individual who somehow managed to combine being both the Lord Chancellor of England and one of the founders of the scientific method.) Above all, Wotton was fascinated by Kepler's camera obscura (described in some detail in Kepler's *Optics*), which he had used to produce a highly accurate drawing of a nearby landscape.

After an outline of various mathematical and musical ideas in the first four volumes, *Harmonices Mundi* culminates in volume five, where Kepler sets out his ideas on the harmonies that he believes exist in the Solar System. His one great triumph in this volume is what has become known as Kepler's Third Law of planetary motion, which states that

The cube of the average distance of a planet from the Sun divided by
the square of the time to complete one orbit is equal to a constant,
which has the same value for every planet.

Unlike Kepler's first two laws, which deal with the way a single planet
moves, the Third Law highlights a relationship between the orbits of the
planets. It is most easily demonstrated if distances are given in terms of
an Earth-Sun distance equal to one unit, and if times are given in terms
of one Earth year. It is worth stressing that (at this stage in the history
of astronomy) nobody had any idea of the actual distances between the
planets and the Sun. All that Copernicus and Kepler could manage were
the relative distances. So if the average distance between the Earth and
the Sun was taken as 1 unit, then they had been able to calculate—for
example—that the Sun-Mars distance was 1.52 units. But the absolute
size of the Earth-Sun distance was an unknown and would remain so for
some 140 years after Kepler's death.

However, absolute distances are not needed to demonstrate the
validity of the law. Using only relative units, the cube of the distance is
the same value as the square of the time, as table 7.1 shows:

Table 7.1. Demonstration of Kepler's Third Law.

	Distance from the Sun	Time to complete one orbit (years)	Distance cubed	Time squared
Mercury	0.387	0.241	**0.058**	**0.058**
Venus	0.723	0.615	**0.378**	**0.378**
Earth	1	1	**1**	**1**
Mars	1.524	1.881	**3.54**	**3.54**
Jupiter	5.203	11.863	**141**	**141**
Saturn	9.537	29.448	**867**	**867**

This very simple and precise law provided one of the best arguments for the Copernican viewpoint. No such law could possibly exist
for the old Ptolemaic system, since it was impossible to be certain of
even relative distances under that system. It also made Tycho Brahe's
compromise system considerably less plausible—after all, why should
the Earth be just like the other planets in obeying the Third Law if

(according to Tycho) it was actually the case that the Sun was going around the Earth? The Third Law makes sense only if the Sun is in some way the cause of the motion of all the planets, including the Earth.[25]

Unlike his first two laws (for which Kepler's explanations in *Astronomia Nova* are painfully lengthy), Kepler has not left any clues as to how he came across this law. Nor (as with his first two laws) did he have any idea why it was the case, other than attributing it to God's desire to create a mathematically harmonious whole. He probably arrived at it simply by a process of trial and error. (If only he had plotted a graph of the logarithm of planetary distances against the logarithm of planetary periods, he would have found it immediately.) He certainly saw it as a vindication of his approach in seeking simple numerical relationships between planetary orbits. (The first step in working out the true reason for the law would require the genius of Isaac Newton some sixty years later.) He records that he first discovered the law on March 8, 1618, but rejected it because of what sounds like an error in his calculations. He then rediscovered it on May 15, 1618.[26]

Critically, and like any good scientific theory, the Third Law could also be used to make predictions. If further planets were to be discovered (as Uranus would be some 150 years later, and Neptune would be some 230 years later), then they, too, should follow this law—as indeed they do.

Unfortunately, the Third Law is a far from prominent feature in volume five of the book. In most of it, Kepler gives full rein to his mystical and numerological outlook and introduces or revisits ideas that we now know are just plain wrong. He still clings tenaciously to his idea of five perfect solids and repeats the most unlikely of justifications for their order. But he had been troubled for some time by the uncomfortable fact that the planetary distances don't quite fit in with those solids.[27] So he returns to an idea dating back to Pythagoras and looks in addition for musical relationships in the ratios of the planetary orbits (where none exists). He then spends far more time on these relationships than on his Third Law. Page after page is filled with tortuous and convoluted arguments trying to explain exactly what the musical relationships are. There is just one brief injection of black humor into his account, when

Kepler uses the musical notes he believes are emitted by the Earth to explain why there is so much suffering on the planet: "The Earth sings MI FA MI, so that even from the syllable you may guess that in this home of ours MIsery and FAmine hold sway."[28]

Kepler could sometimes be utterly wrong, just as he was often brilliantly right. In volume five of *Harmonices Mundi* (just as with his five perfect solids), he has been completely led astray by his belief in a deity who had constructed the Solar System along strict geometrical and musical lines. Because of this belief, he was once again falling into the trap of seeing patterns in nature where none existed. His oft-quoted hymn of praise in this book that "I am stealing the golden vessels of the Egyptians to build a tabernacle to my God"[29] was totally inappropriate, because the tabernacle he was building in the book (apart from the Third Law) would turn out to be worthless. And whereas his five solids were the inspiration for his later achievements, these later ideas were nothing but an unfortunate cul-de-sac. If *Harmonices Mundi* had been his only work, later generations would have seen him as an eccentric and idiosyncratic character who had come across the Third Law entirely by accident.[30]

Nowadays, we can divide Kepler's work neatly into the "scientific," where he made huge advances, and the "mystical," where we can see that he was hopelessly wrong. But for Kepler, these two aspects of his work were always an integrated whole. Throughout his life, he thought he was taking a single unified approach to nature, which he saw as the creation of his god—a mathematically motivated deity who bore a remarkable resemblance to Kepler himself.

EPITOME ASTRONOMIAE COPERNICANAE

Much the same conclusions can be reached over Kepler's *Epitome of Copernican Astronomy*. This was published in installments in the years 1618–1621. Kepler was too modest to give it the more accurate title of *Epitome of Keplerian Astronomy*. Book four starts with a welcome admission that the key conclusions in his earlier *Astronomia Nova* had been "hidden . . . in thickets of calculations."[31] The *Epitome* tries to remedy this by producing a summary "in plain and simple speech,"[32] and is in the form of questions and answers about Kepler's views on astronomy. What follows is an inextricably intertwined mixture of scientific correctness and mystical error.

Kepler restates his Copernican belief that the Sun is fixed and immovable at the very center of the Universe. For him, the apparent threefold nature of the Universe (the Sun, the outermost celestial sphere to which the fixed stars were attached, and the space in between) was a representation of the Christian Trinity—God the Father, God the Son, and God the Holy Spirit. The Sun, he thinks, is not at all the same as the fixed stars. The gradual realization that our Sun is in fact just one of hundreds of billions of stars in our Milky Way Galaxy, and that it is not in any way special, would not come until long after Kepler's death. Once again, his mystical and religious beliefs were leading him astray.

He spends more time on his five perfect solids, again trying to justify their significance as the basis for the construction of the Solar System. He again tries to find a pattern where none exists in his claim that the diameters of the planets grow larger in proportion to their distances from the Sun—although we now know that this is not so. (Mars, for example, is smaller than the Earth, and Saturn, Uranus, and Neptune are smaller than Jupiter.) He also maintains that the Sun is the densest body in the Solar System, and that the density of the planets declines with distance from the Sun. In reality, the Sun (which is composed of ionized gas at extremely high temperatures) has a low density in comparison with the rocky planets Mercury, Venus, Earth, and Mars.

In between these erroneous speculations, some of the correct conclusions of his earlier writings reappear. Kepler repeats his critical understanding that the Sun is responsible for the force that moves the planets. He was the first person ever, in *Mysterium Cosmographicum*, to make this claim, and it was ultimately the basis for his later discoveries. But he still wrongly attributes the force to the Sun's rotation, rather than to its gravitational field—the latter concept would not arrive until Newton. He also restates his key discovery, set out in *Astronomia Nova*, that the planets go around the Sun in ellipses, and he gives a clearer statement of his Second Law.[33]

The main saving grace of the book is that it repeats Kepler's Third Law and then extends it to include the four newly discovered satellites of Jupiter. The distances and periods that Kepler quotes for these satellites were not totally accurate (unsurprisingly, since observations of them were at an early stage), but his Third Law held up well, given these inevitable inaccuracies.[34] The value of the cube of the distance from Jupiter divided by the square of the time to go once around Jupiter should be exactly the same for all four satellites. Using Kepler's values, they come very close to being the same, and using modern values they are identical.

It is in this book that he acknowledges his huge debt to three people. "I build my whole astronomy upon Copernicus's hypotheses concerning the world, upon the observations of Tycho Brahe, and lastly upon the Englishman William Gilbert's philosophy of magnetism."[35] Kepler uses William Gilbert's explanation of magnetic action at a distance to justify by analogy the action at a distance of the Sun on the planets. He greatly admired Gilbert and had once said about him that "[he] appears to have made good what was lacking in my arguments on Copernicus's behalf through his admirable skill and his industry in collecting observations on the study of magnets."[36]

REPRINT OF *MYSTERIUM COSMOGRAPHICUM* *(THE MYSTERY OF THE UNIVERSE)*

The year 1621 also saw the reissue of *Mysterium Cosmographicum*, twenty-five years after the first edition. Although this second edition contained plentiful footnotes, containing corrections to—or explanations of—the original version, Kepler showed no sign of wanting to depart from his central thesis: that the spacing between the planets was primarily determined by the five regular solids. On the contrary, the second edition proclaimed that the whole of his astronomical output since then directly related to one or another of the chapters in *Mysterium Cosmographicum* and was either an illustration of something in the book or a completion of it.[37] Not merely that, but his whole life and work had been as a result of this one little book.[38]

KEPLER'S MOTHER: THE WITCH TRIAL

"Thou shalt not permit a witch to live." This verse from the Bible[39] made it clear not only that witches existed but also that they had to be put to death. The flames of hysteria were fanned after the publication in 1487 of *Malleus Maleficarum* (*The Hammer of Witches*), by Heinrich Kramer, a Dominican priest and witch hunter. Here at last was something that Protestants and Catholics could agree on. It is estimated that, in the period 1450–1750, something between forty thousand and one hundred thousand people throughout Europe were executed after being convicted of witchcraft.

Torture was found to be remarkably effective in extracting confessions of sexual liaisons with the devil, and it consisted of the application of thumbscrews, followed by stretching on a rack. The toll fell mainly on older and poorer women. The practice only gradually and sporadically came to an end with the coming of the Enlightenment. Witch hunts and subsequent executions were conducted in a patchy way throughout Europe. In Spain and Italy, England and Scotland there were relatively few. In the German states, executions were disproportionately higher, and at least twenty thousand witches were put to death.[40]

Kepler's mother, Katharina, was very nearly one of them. If her son had not still held the title of Imperial Mathematician to the Holy Roman Emperor, and had not thrown himself enthusiastically into her defense, it is unlikely that she would have survived. In Weil der Stadt, the town of Kepler's birth, thirty-eight witches were executed in the period 1615–1629. In Leonberg, where Kepler's mother still lived, six women had been condemned as witches just in the two years 1615–1616.[41]

Kepler himself undoubtedly believed in the existence of witches, as was the norm for that time. Explicit passing references to witches can be found in his *Somnium*[42] and in his *Optics*,[43] and both apparently take their existence for granted. But it is equally certain that he did not think that his own mother was a witch.

From Kepler's earlier descriptions of her, we know that when she was younger his mother could be a stubborn, unpleasant, and quarrelsome person. And, at the age of sixty-seven, after a hard life during which she had borne seven children (three of whom had not survived to adulthood), her appearance must also have been against her. (It was well known that ugliness was a characteristic of witches.) Matters had begun in August 1615 with a quarrel between one Ursula Reinbold of Leonberg and Kepler's mother and brother Christoph, who both said unpleasant things to her face about Frau Reinbold's alleged activities as a prostitute.

Frau Reinbold—just like Frau Kepler—was not one to forgive an offense, and she spread the rumor that Frau Kepler had tried to poison her, using a witch's potion. Frau Kepler was not much liked, and this rumor led to other accusations being made. A close relative of hers, in whose home she had lived, had been burned at the stake as a witch. She had tried to have her late father's skull mounted in silver and converted into a drinking cup. She had ridden a calf to death. The local schoolmaster who used to read Kepler's letters out loud to her had been given a drink by her and had suffered pains and had eventually become lame as a result. When Frau Kepler had walked past the local butcher, he had felt a sharp pain in his thigh, even though she had not touched him. She had caused the death of the local tailor's two children. She had walked through locked doors. The case against her was indeed damning.[44]

In a critical development, in front of the local magistrate, Ursula Reinbold's brother threatened Frau Kepler with violence, at sword point, if she didn't reverse a spell she had placed on his sister. Fortunately she refused: to have done otherwise would have been a clear indication of guilt. Frau Kepler (assisted by her son Christoph, her daughter Margarete, and her son-in-law, a pastor, Georg Binder) then tried to bring a complaint to the local court, but this was constantly delayed, apparently with the active connivance of the magistrate.[45]

Kepler first heard of his mother's plight in a letter that arrived on December 29, 1615. Four days later, he fired off an angry letter to an official in Leonberg, protesting strongly about the treatment his mother had received. Little then happened until October 1616, when a dramatic event made an eventual witch trial inevitable. Frau Kepler passed by a group of girls in a field, brushing against their clothing. One of the girls later stated that Frau Kepler had hit her arm. The consequence, she asserted, was that her hand had become paralyzed. The local magistrate was quickly able to confirm that the bruise on her arm had certainly been made by a witch. (Although, in spite of this, the girl recovered within a few days.) But then Frau Kepler made a big mistake—she offered the magistrate a silver cup if he would drop his report on the incident. This not only didn't work; it was taken as further evidence of her guilt. In a panic, she then compounded her error by fleeing to Linz, arriving at Kepler's home in December 1616. The authorities back in Württemberg agreed to arrest Frau Kepler on sight and to begin a witch trial, but they could do little while she remained outside their territory.[46]

Frau Kepler remained with her son and his family for some nine months. She was there in July 1617 for the birth of Katharina—named after her grandmother—and for the death in September of her granddaughter Margareta. She finally left in October 1617, to return to her daughter, who lived close to Leonberg. Her departure coincided with the arrival of the news that Kepler's much loved stepdaughter, Regina, had died at the tragically young age of twenty-seven.[47] Kepler followed his mother back to Württemberg. On the way, at the request of Regina's widower, he deposited his daughter Susanna in Walderbach (near Regensburg), where Regina had lived, so that Susanna could look after Regina's three children. He then returned to Linz in December.

The process of gathering evidence against Frau Kepler dragged on for a couple more years. Eventually, on July 24, 1620, her arrest was ordered. In the event that she were to plead not guilty to the charges of witchcraft, she was to be tortured until she changed her plea. She was arrested on the night of August 7 at her daughter's house, carried away in a wooden chest (to avoid publicity), and imprisoned, initially in Leonberg. But even with the threat of torture hanging over her, Frau Kepler still refused to confess. In September 1620, Kepler left Linz for Württemberg, with the aim of defending his mother at the trial. At this and every stage in the process, Kepler showed complete loyalty to her. His brother Christoph and his brother-in-law, the pastor, Georg Binder were much more lukewarm in their support. (After all, they did have their reputations to consider.)[48]

The trial lasted for a full year, and poor Kepler "was involved for the whole year in my mother's case."[49] At one point, Frau Kepler had failed to cry when, on the reading of certain documents, she was expected to show remorse. In explanation, she pointed out that she had cried so much in her life that she could not cry anymore. Kepler's active role in the case caused one of the scribes at the trial to report that "the prisoner appears, alas, with the support of her son, Johannes Kepler, the mathematician."[50]

Eventually, the papers were sent to the law faculty at Tübingen University for a decision. Their conclusion was that Frau Kepler should be shown the instruments of torture, with a view to persuading her to confess. In September 1621, she was taken to the torture chamber. Under enormous pressure, she continued to insist on her innocence. Following this, the Duke of Württemberg finally had her pardoned and released. But given all she had been through, it was hardly surprising that this seventy-four-year-old woman died in April of the following year, barely six months later.[51]

PORTRAITS

In 1620, before leaving for his mother's trial in Württemberg, Kepler had commissioned a portrait of himself (by an unknown artist), which he sent to his friend Matthias Bernegger in Strasbourg. The likeness was not a very good one, but Bernegger nevertheless donated it to Strasbourg University, where it still hangs in the library. A copy of the painting is kept in the Kepler Museum in Regensburg.

Bernegger went on to have a copper-plate engraving (shown on the front cover of this book) prepared from the portrait so that an unlimited number of images could be made of Kepler. While it was no doubt done with the best of intentions and clearly showed a growing public interest in Kepler, it was an even worse likeness. Kepler's friends in Tübingen certainly didn't think much of it. One of them, Thomas Lansius, with tongue firmly in cheek, composed a poem that jokingly provided a pseudo-astronomical explanation of why it was that the engraver (Jacob van der Heyden) had got it so badly wrong:

> It bears Kepler's name, but the image is entirely wrong.
> But tell me, why does the artist make such a mistake?
> The problem is the moving Earth, which moves according to
> Kepler's Laws.
> The force of the rotation also led the painting hand astray.
> If the Earth did not move, but always stayed still,
> The picture of Kepler would not be so badly distorted.[52]

Image 7.2. The portrait at the Kremsmünster science museum.

So it is unfortunate that the most frequently reproduced image of Kepler is not a good likeness. There are only two other possible contenders. One is a (now slightly damaged) portrait that can be found in the astronomy section on the top floor of the science museum at Kremsmünster (near Linz). It is frequently reproduced, but its authenticity is uncertain. The other is a work believed to be a portrait of Kepler by the German mannerist painter Hans von Aachen (1552–1615), who became Rudolph's official court painter, and whose time in Prague largely overlapped with that of Kepler. This painting is now displayed in the castle gallery at Rychnov nad Kněžnou, in the Czech Republic.

Image 7.3. The portrait believed to be by Hans von Aachen.

THE THIRTY YEARS' WAR (1618–1648)

There is a certain irony in the fact that Kepler discovered his Third Law, showing the harmony that governed the movements of the planets, just eight days before a total lack of harmony broke out on Earth.

The Thirty Years' War was one of the deadliest conflicts ever to engulf Europe. It was responsible for some eight million deaths, over one-third of the population of Germany at the time.[53] Its causes were complex but—at least in its initial stages—were largely religious. The spark that ignited it took place in Kepler's former home city of Prague (which he had wisely left), and is now known as the Defenestration of Prague.[54] On May 23, 1618, in Prague Castle, a small group of Protestant noblemen of Bohemia threw three senior figures in the (Catholic) Habsburg administration out of a high window. Miraculously, they survived the fall. Some said this was because they had been rescued by the direct intervention of the Virgin Mary.[55] Others said their survival had more to do with the pile of horse dung that had been fortuitously left just below the window.[56] The incident quickly led to armed conflict that spread over much of the Holy Roman Empire and beyond and lasted for the next thirty years.

Rudolph's younger brother Matthias died in March 1619. His (and Rudolph's) fiercely Catholic cousin Ferdinand (whom Kepler had encountered in Graz) was soon elected the new Holy Roman Emperor. Only two days earlier, the Protestant Frederick V (son-in-law of James I of England) had been elected by the Protestants as the new king of Bohemia, so setting the scene for conflict between the two. Frederick's reign lasted just one year. His army was defeated (at the battle of the White Mountain on November 8, 1620) by the Catholics under the overall control of Duke Maximilian of Bavaria. Maximilian was another Jesuit-educated and devout Catholic, like his brother-in-law Ferdinand. (Because of the incestuous nature of Habsburg marriages, Maximilian managed at one stage to be both the brother-in-law and the son-in-law of Ferdinand.) Prague was retaken later that day, just after Frederick had ignominiously fled the city.

A couple of dozen leading Protestants in Prague were arrested and executed on the direct orders of Ferdinand. One of these was Kepler's great friend Johannes Jessenius, who had acted as the neutral arbiter in Kepler's dispute with Tycho (and who had given Kepler useful information on the structure of the eye for his book *Optics*). His grisly beheading took place in June 1621 in the center of Prague, in front of Tyn Church (where Jessenius had given his eulogy to Tycho Brahe twenty years earlier), but not before his heretical tongue had been brutally ripped out and nailed to the scaffold.[57]

Duke Maximilian had also conquered Upper Austria, including its capital, Linz, in July 1620, a couple of months before Kepler left Linz for his mother's trial . After the trial came to an end, and Kepler returned to Linz in November 1621 (to a house that can still be visited at 5 Rathausgasse, close to the town's main square), Maximilian's army was still firmly established in the city.

Kepler's main task now was the completion of the *Rudolphine Tables*. His First Law had established the shape of each planet's orbit, and his Second Law had established the position of the planet in this orbit at any moment in time. But it was still necessary to make use of these two laws to perform a complicated series of calculations to determine where the planet would appear in the sky at any time, as seen from Earth. The calculations would result in the *Rudolphine Tables*, which would achieve exactly this. Their accuracy would provide one of the best arguments for the fundamental truth of the Copernican position. But their construction required considerable work. This was not a task he relished because of the mass of tedious calculations it involved. He had pleaded: "Don't sentence me completely to the treadmill of mathematical calculation and leave me time for philosophical speculations, which are my sole delight."[58] And he was now potentially in the same position in Linz as he had been in Graz. A fanatically devout Catholic (this time Maximilian, rather than Ferdinand) was beginning to oppress the largely Protestant population.

The effort of calculation was eased by newly discovered logarithms (in 1614, by the Scottish mathematician John Napier). By converting lengthy and tedious multiplications into simple additions, logarithm tables reduced enormously the burden of calculation. Kepler took full advantage. In complete contrast, Michael Maestlin's grumpy reaction to logarithm tables was rather like the attitude of many in the twentieth century when pocket calculators became available: "It is not seemly for a professor of mathematics to be childishly pleased about any shortening of calculations."[59]

The tables were completed in 1624, but the printing press in Linz was not up to the complex task of printing them. He would have to move elsewhere. But where? The Thirty Years' War was in the process of devastating large parts of Germany, which explains Kepler's dry comment: "Which place should I choose—one already laid waste or one which has yet to be laid waste?"[60] One possibility was the Free Imperial City of Ulm, a bastion of Lutheranism. Not surprisingly, Ferdinand refused to give permission for this. It looked as though the printing would have to be done in Linz after all.

Then in 1625, the Catholic rulers of Linz began a drive to expel the Protestant preachers and teachers, just as they had done in Graz some twenty-five years earlier. Kepler—now far more famous than he had been when in Graz, and also responsible for what was widely seen as the critical task of producing the *Rudolphine Tables*—found that he was to a large extent immune from the general persecution:

> It is a great comfort that we are not burned, but are allowed to live, if there is any meaning in the permission to live for him who has been deprived of the necessary means of life. . . . But this does not apply to me. For, so far, I have been allowed the privileges of a courtier.[61]

They also took steps to get rid of any heretical books in the town. This time, Kepler was not immune. Anyone with a love of books will know how he felt:

> My whole library has been sealed up since 1st January [1626], with the exception of a very few books. To regain the books, they set the condition that I myself should select those which are to be surrendered; that means that the bitch must surrender one of her young ones. The mark of such slavery burns.[62]

The Catholics' harsh rule of Upper Austria, and in particular their attempts to enforce conversion to Catholicism, proved too much for many. In May 1626, a peasants' revolt began, and their makeshift army laid siege to Linz in June. The siege lasted for a couple of months before the emperor's army was able to break it. The revolt itself was only finally—and brutally—put down by Maximilian in November.[63] One consequence of the siege, however, was the destruction in a fire of the printing press that Kepler had been using for the *Rudolphine Tables*. It would no longer be possible to produce them in Linz. The hard time Kepler was having was made clear in a classic understatement of his. He wrote in October to his friend the Jesuit Paul Guldin, saying that "it is an extremely strange fate which permanently handicaps me. Again and again, difficulties arise through no fault of mine."[64]

In a later letter to Matthias Bernegger, he wrote:

As you wish to know how I am, you may be glad to hear that—with the help of God and his angels—I safely survived the siege for 14 days. I did not starve either, even without tasting horsemeat. Only a few had the same good luck. When with the approach of the imperial troops the siege became less tight, I sent a petition to the court in which I asked for permission to travel to Ulm.[65]

At last, Kepler had got his way. Ferdinand no longer had any choice in the matter. Permission was given, and the moment had finally come to leave Linz.

CHAPTER 8

THE FINAL YEARS (1626–1630)—
THE *RUDOLPHINE TABLES*

K epler, Susanna, and their three surviving children (Cordula, Friedmar, and Hildebert) finally left Linz for Ulm in November 1626.[1] The Danube, Europe's second-longest river, conveniently meanders through both Linz and Ulm as it winds its way through central Europe, and it provides a handy means of traveling between the two cities. The family was able to use this route as far as Regensburg.[2] Here they stopped for a while, and Kepler found lodgings for his wife and children at a house (now located at 2 Keplerstrasse) in the old town.[3] The remaining stretch of the Danube was too icebound at that time of year, so he completed the journey by road. On arriving in Ulm, he stayed in a house at 3 Rabengasse,[4] a street in the shadow of the huge Gothic minster that dominated the town. It was also conveniently close to Ulm's printing press.

WEIGHTS AND MEASURES

The main task was now to supervise the printing of the *Rudolphine Tables*. But the town council took the opportunity to make use of its distinguished visitor for another undertaking. The system of weights and measures in Ulm at that time was in a complete mess. What were meant to be definitive measurements of weights and volumes differed from each other by significant amounts and provided ample opportunity for fraud. Kepler was asked to devise a replacement system.

He recommended the creation of a standard weight, standard length, and standard volume, all with reference to a single cylindrical vessel. He contemplated moving to a decimal system, but this would have been too great a departure from the units already in existence. So he stuck as closely as possible to existing measures.

Image 8.1. The huge Gothic minster in Ulm.

JOHAÑES
KEPPLER
ASTRONOM ☆1571 † 1630.
K·gab hier 1627 d·Rudolfiniſchen
Tafeln heraus & ſchuf durch d·ſg·
Kepplerkeſſel d·Grundlage z·
einem geordnet·reichsſtädti-
ſchen Maſs-&Gewichts -
weſen

VEREIN F. KUNST U·ALTERTUM § VEREIN ALT-ULM
STADTGEMEINDE ULM

Image 8.2. A plaque commemorating both Kepler's stay and his achievements in Ulm.

The starting point was the Ulm hundredweight (cwt), previously defined in terms of the weight of a set number of coins (and unfortunately these could vary in weight). Kepler realized that if he were to use a vessel of a convenient size, holding about 3.5 cwt (about 390 lbs.) of water (on the old coin-related definition), he could redefine 3.5

cwt as the weight of water in the vessel. He could then define the diameter of the vessel as one Ulm yard (= 0.6 meters in modern units) and the height (in rather oddly named units of length) as 2 Ulm shoes (= 0.584 meters), so providing standard lengths. The volume of anything (typically either wine or grain) held by the vessel (= 164.6 liters) then also provided a standard unit of volume.

The task of manufacturing the vessel was given to Hans Braun, who was in charge of the Ulm foundry. It was cast in bronze, according to Kepler's typically precise specifications. For decoration, and to help carry the vessel, the heads of four predatory birds—also in bronze—stuck out from around the rim of the vessel. Around the base, to give stability, were four bird claws. Kepler was a frequent visitor to supervise production. The cast was ready for his inspection on November 9, 1627, but—perfectionist as ever—he criticized it as not being sufficiently accurate. He had to leave Ulm a couple of weeks later, so the supervision was taken over by Johannes Faulhaber. The final product was duly used to provide Ulm's standard weights and measures. The Kepler Kessel, as it is now known, can still be seen in the museum in Ulm.[5]

THE *RUDOLPHINE TABLES*

Kepler had to juggle his responsibilities for the new weights and measures system with the supervision of the printing of the *Rudolphine Tables*. The chosen printer was named Saur.[6] His surname was an appropriate one, because he turned out to be a very difficult person to work with (*Säure* is German for "sourness" or "acidity")—or, perhaps more likely, he simply found Kepler unreasonably demanding in what he required. At one stage, in February 1627, Kepler left Ulm in disgust, hoping to find an alternative printer in Tübingen. But the winter weather combined with his health problems forced him to turn back and make his peace with the recalcitrant printer.[7]

The culmination of Kepler's life's work finally came in September 1627 when the printing was completed.[8] The tables, based on Kepler's first two laws of planetary motion, enabled the calculation of the most accurate forecasts of planetary positions (by a huge margin) that had ever been achieved. Their unsurpassed accuracy would provide enormous support for the Copernican stance and would also lead to the gradual—and no doubt reluctant—acceptance of Kepler's ellipses. It seems entirely fitting that this vital work should have been published in Ulm, which was to be the birthplace of Albert Einstein just over 250 years later.

Figure 8.1. Frontispiece to the *Rudolphine Tables*.

The frontispiece to the tables (shown in figure 8.1) was drawn up according to Kepler's instructions and is a beautiful and intricate summary in a single picture of the key people and publications in astronomy over the previous two thousand years.[9] In the center of a temple is a gathering of five astronomers—a Babylonian in the background, the ancient Greek astronomers Hipparchus (with a copy of his star catalog) and Ptolemy (with a copy of his crucial astronomical work the *Almagest*), and also Copernicus and Tycho. Copernicus is seated, with his book *De Revolutionibus* on his lap. Tycho, standing, points to a diagram of the Solar System on the ceiling, while saying, "Quid si sic?" ("What if it's actually like this?").

On the base, on the left-hand panel, is a picture of Kepler, working away by candlelight (as he must have done on numerous occasions). Above him are the titles of the four other key works that he hoped he would be remembered for: *The Mystery of the Universe* (showing that, even now, he had not abandoned the idea that the geometry of the solar system was determined primarily by the five perfect solids), *Optics*, *The New Astronomy* (still modestly labeled as the *Commentaries on Mars*), and the *Epitome of Copernican Astronomy*. The central panel on the base is a map of the island of Hven, from which Tycho had made most of his observations, and the right-hand panel shows Georg Celer, who engraved the illustration, with his assistant.

On the very top of the temple sits the goddess Urania, the muse of astronomy, holding up a victor's wreath (perhaps intended for Kepler?). Over her hovers a crowned eagle, the symbol of the Holy Roman Emperor (now Ferdinand), dropping coins, no doubt alluding to the fact that poor Kepler was still owed substantial sums of money for his efforts. A few of the coins have landed on Kepler's table. Six other goddesses stand precariously on the roof of the temple, each one illustrating a different aspect of Kepler's work. The goddess who is third from the left provides a good illustration of the degree of detail in the picture. Her halo[10] is made up of the numbers 6931472, which are the first seven decimal places of the natural logarithm of the number 2. This was entirely appropriate, as Kepler had made great use of the newly discovered logarithms to carry out his calculations.

The title page prominently displays Tycho's name. This was not just because Tycho's relatives were still causing problems. It was also because Kepler remembered Tycho with genuine gratitude and affection.

In his preface to the tables, Kepler draws attention to the vast improvement in accuracy that he has achieved. Prior to the *Rudolphine Tables*, people had made use of two other sets of tables. The first was the *Alphonsine Tables* (named after Alphonso X, the thirteenth-century king of Castile mentioned in the introduction). These were based on the Ptolemaic system and had been largely superseded by the *Prutenic Tables*. The *Prutenic Tables* were based on the Copernican system, although their author—Erasmus Reinhold—had taken care not to mention Copernicus's absurd idea of a moving Earth. Even these tables, Kepler points out, lacked accuracy:

> For example, throughout the whole year [1625], Mars has been observed much farther advanced in the sky than the Prutenic calculations predicted, and the error has grown through the months of August, September and October to the magnitude of four, and almost five, degrees.[11]

He also laments the fact that the tables had been so long delayed, "especially due to the intervention of wars."[12] And he reminds readers of perhaps his greatest accomplishment, which today we take for granted:

> the unexpected transfer of the whole of astronomy from artificial circles to natural causes.[13]

Kepler went to Frankfurt to make arrangements for the sale of the book and finally returned to his family in Regensburg, via Ulm, in November.[14] December 1627 was crunch time; what was he to do next? He traveled on to Prague to the court of the Emperor Ferdinand II, to present him with a copy of the *Rudolphine Tables* and to find out about the prospects for a new job. He still held the title of Imperial Mathematician, and he was still the district mathematician for Upper Austria, but religious problems there meant that a return to Linz was out of the question. (He had already put out feelers to his friend Bernegger, who had valiantly looked around for a suitable post, but these efforts had come to nothing.[15])

Somewhat to Kepler's surprise, Ferdinand and others at his court were favorably impressed by his *Rudolphine Tables*. He was offered payment of 4,000 gulden for his work—but the catch was that this money had to be paid by the towns of Ulm and Nuremberg, rather than directly by the emperor. He was even offered a prestigious post, but only if he would convert to Catholicism. So he bravely rejected the offer.

SAGAN

His consolation prize was to work for Albrecht von Wallenstein, the man for whom he had written a horoscope over twenty years previously, and who was now the very successful commander of the emperor's army. Wallenstein had just been given Sagan in Silesia, and he offered Kepler a post there, together with accommodation, a salary, and his own printing press. So in May 1628, Kepler left Prague for Regensburg to fetch his family, then paid a brief final visit to Linz to wind up his affairs there. He and his family finally arrived in Sagan late in July 1628.[16]

Kepler suffered from a dreadful sense of isolation in Sagan, which was far from his previous homes and was hardly a great intellectual center:

> For it is loneliness which makes me feel oppressed here, far away from the large cities; and letters come and go only slowly.[17]

One advantage of the town was that its inhabitants were almost entirely Protestant. But even this was to change. In November 1628, shortly after his arrival, the order went out that they were either to convert to Catholicism or leave—although, once again, Kepler was exempt from this directive.[18] He threw himself into the work of preparing Tycho's observations for publication, preparing ephemerides (tables of values—in this case, based on the *Rudolphine Tables*—that give the positions of the planets at any given time), and continuing with his writing of *Somnium*, but it took a further year before his printing press was finally set up.[19]

At one stage, Wallenstein wanted Kepler to move from Sagan in order to take up the post of professor of mathematics at the University of Rostock, in the state of Mecklenburg-Schwerin, in the far north of Germany. Kepler had no desire at all to travel to such a cold and distant region. Besides, the people there were so uncivilized. It would have been even worse than Sagan. Matters were further complicated by the fact that Ferdinand had just dismissed Wallenstein from his post as army commander. Fortunately, Wallenstein's idea was eventually dropped.[20]

JACOB BARTSCH

Jacob Bartsch had first appeared in Kepler's life when the two met briefly in Ulm. Bartsch had been a student of astronomy and medicine in Strasbourg. The two seem to have got on well. When the *Rudolphine Tables* appeared, Bartsch used them to produce his own calculations of ephemerides and offered Kepler further help. Kepler seems to have been able to afford an assistant, and in 1628 he took him on to carry out further tedious calculations. At some stage, it occurred to him that Bartsch might make a good match for his daughter Susanna, by now in her midtwenties and still unmarried. Susanna was now living and working some distance away, in the small town of Durlach, to the west of Stuttgart (and a very long way from Sagan), but she was relatively close to Strasbourg, where his old friend Bernegger lived. In April 1629, he wrote to Bernegger to try to find out more about Bartsch:

> The gentleman in question (Jacob Bartsch) lives in my vicinity and helps me with my arithmetic. He is still a bachelor, and has postponed marriage. . . . Please try to find out how he has lived in Strasbourg, what habits he showed, how much money he used, how much hope can be put on his getting a professorship in Strasbourg.[21]

The response must have been satisfactory because when Bartsch asked Kepler for permission to marry his daughter (before he had even met her), Kepler willingly agreed—on condition that Susanna also agreed, which she did. Susanna was married to Bartsch on March 12, 1630, in the magnificent Gothic cathedral in Strasbourg.[22] Distance and the pregnancy of Kepler's wife prevented them both from being present for the wedding, but his loyal friend Bernegger gave him a glowing description of the event:

> I myself, according to your wishes, have in a way substituted for you, supported by your excellent brother Christoph and your son Ludwig. Protection and ornament to the bride in attending her were your sister Margarete and your cousin, the wife of Dr Marchtrencher, ... All in all, the bridal procession was made up of notable men and women of all ranks, the elite of the whole town. Consequently, I have seldom seen so many people gathered together. Everywhere the streets through which we went were crowded. The large number of spectators who streamed in might have filled quite a large town. But do not think for a moment that this honor was given only to the bride and groom. It was meant especially to honor you. ... I congratulate you with all my heart on such a son-in-law and such a daughter. You could not wish for better ones.[23]

One month later, Kepler's twelfth—and last—child, Anna Maria, was born.

KEPLER'S DEATH

Probably to try to recover the large sums of money that were owed to him, in October 1630 Kepler set out for Regensburg, where Ferdinand was meeting with his congress of electors. He arrived in Leipzig on October 14, where he stayed with his friend Philip Muller ("my second Bernegger"). From here, he traveled on to Regensburg, which he reached on November 2. He had made his final journey on a skinny mare, which he then sold for 2 gulden.

He stayed at the house of a friend, at what is now 5 Keplerstrasse, a very short distance from where his family had previously lodged. Three days later, he developed a fever (perhaps a recurrence of the same quartan fever that had troubled him intermittently for much of his life). The fever worsened, and he became delirious. Some Lutheran priests came to see him to try to comfort him. He died on November 15.[24]

Image 8.3. Strasbourg Cathedral.

Even in death, religious strife continued to follow him. Regensburg was a Free Imperial City that had opted to become Lutheran.
However, the Catholic minority had held on to the cathedral and also

to the cemetery that was inside the city walls. It was clearly out of the question for Lutherans to be buried in a Catholic burial ground, so the Lutherans had created two cemeteries of their own. Lack of space had meant that these had to be located outside the city walls. Kepler was buried in one of them.[25]

Two years later, it was destroyed—along with Kepler's grave—in the Thirty Years' War. All we are left with is the inscription that Kepler had composed himself and had asked to be placed on his tombstone:

> Mensus eram coelos, nunc terrae metior umbras.
> Mens coelestis erat, corporis umbra iacet.[26]
> (I measured the skies, now the shadows of the earth I measure.
> Sky-bound was the mind, the shadow of the body lies here.)

CHAPTER 9

EPILOGUE—
THE REAL UNIVERSE

Although the Regensburg cemetery containing Kepler's grave was destroyed, there is now a monument to him in the park that borders Maximilianstrasse, on the southern outskirts of the old town and a short distance from the original location of the cemetery. The house at 5 Keplerstrasse, in which he died, has been restored to its original state and is now a Kepler Museum.

Michael Maestlin, Kepler's mentor and friend, and his senior by over twenty years, died in 1631 at the ripe old age (certainly for that time) of eighty-one. Also in 1631, the French astronomer Pierre Gassendi observed the transit of Mercury across the face of the Sun that was predicted in Kepler's *Rudolphine Tables*. This was the first time that such a transit had been seen, and its successful prediction added considerable weight both to the Copernican stance and to Kepler's laws, on which the tables had been based. Gassendi wrote that "men like Kepler should never die—or like demi-gods, they should live for centuries to come."[1]

The promising career of Kepler's new son-in-law, Jacob Bartsch, came to a sudden and tragic end when he died of the plague in 1633, at the age of thirty-three.[2] His wife, Susanna, Kepler's daughter, married again and went on to live into her sixties. Two more of Kepler's own children, Friedmar and Hildebert, died tragically young—Friedmar in 1633 and Hildebert in 1635.[3]

Kepler's only surviving son, Ludwig, completed the task started by Bartsch of getting Kepler's *Somnium* published, in 1634.[4] It seems unlikely that it raised much money for the family. Kepler's second wife, Susanna, died a few years later, in her late forties.[5] Albrecht von Wallenstein, Kepler's last employer, was assassinated in 1634, probably on the orders of the Emperor Ferdinand.

Image 9.1. Kepler's monument in Regensburg.

In Italy, in 1633, Galileo was put on trial by the Roman Catholic Church for daring to publish a book (*Dialogue on the Two Great World Systems*) in which he argued that Copernicus was correct. Numerous accounts have been written on this chapter in the history of astronomy, some of them even trying to defend the action taken by the Church. However, the bottom line is that an old man of almost seventy years, frail and in poor health, was threatened with torture by the Church simply for putting forward an alternative cosmology. He was forced into a public recantation of his Copernican views, and he was then held under house arrest for the remaining eight years of his life. And his book was considered so dangerous that the Church banned it for the next two hundred years.

Shock waves from Galileo's condemnation spread all over Europe. The French philosopher René Descartes (1596–1650) was on the point of publishing an explicitly Copernican work (*Le Monde*) but withdrew it when he heard about Galileo's fate. Yet thanks to a combination of Galileo's efforts and those of Kepler, within a generation most educated people throughout Europe came to accept that the Earth did indeed go around the Sun.

Annoyingly, Galileo's book makes no mention of Kepler's ellipses, even though Galileo must have been fully aware of Kepler's work in this area.[6] But Galileo, for all his great physical understanding of the laws of motion on the surface of the Earth, was yet another person who still clung to the idea of circular motion in the heavens. He probably thought that the idea of elliptical motion was as strange and unlikely as Kepler's five perfect solids. (Ironically though, it was Galileo who introduced the parabola, another conic section, into terrestrial physics.)[7]

In 1651, the Jesuit astronomer Giambattista Riccioli (1598–1671) provided Kepler with a form of immortality. He produced a map of the Moon that named many of its craters after ancient Greeks and famous people of his own time (including himself).[8] As a good Catholic who disagreed with the idea that the Earth was not at the center of the Universe, he named the most prominent lunar crater after Tycho, in honor of his compromise system. However, he did also name one large crater after Copernicus and a somewhat smaller crater after Kepler. Perhaps not surprisingly, Galileo has no crater of any size or significance named after him. Riccioli's names are still used in modern lunar maps.

Image 9.2. The Moon at last quarter. The craters Copernicus and Kepler can be clearly seen (Credit: Paul Hutchinson, FRAS).

ISAAC NEWTON

Isaac Newton (1642–1727) was born on Christmas Day 1642 (as measured by the old Julian calendar, still in use at that time in Protestant England), in the same year in which Galileo died. Newton was a genius who (with the possible exception of Charles Darwin) is arguably the greatest scientist England has ever produced. He announced his universal theory of gravitation to the world in his *Principia*, published in 1687.

The book accepts as self-evident Kepler's fundamental idea (much disputed in Kepler's lifetime) that there is a physical cause behind planetary motion. Using his theory of gravitation, he was able to deduce mathematically Kepler's three laws of planetary motion. So what Kepler had thought of as three separate and unrelated laws were in fact merely different manifestations of a single very simple law. And from the moment of Newton's discovery, Kepler's name was guaranteed immortality. Newton himself, in a typically mean-spirited way, did not even mention Kepler's discovery of elliptical motion in *Principia*, but there is no doubt that he knew about it.

However, Newton did appreciate that Kepler's laws operated only under idealized conditions, in which the masses of the planets are neglected. For example, it is not strictly true that a planet moves around a fixed Sun in an ellipse. What actually happens is that the planet and the Sun both move in ellipses around their mutual center of gravity. It is simply that the Sun is so massive (it contains about 99.9 percent of the total mass of the Solar System) that the mutual center of gravity of the Sun and the planet is close to or inside the surface of the Sun. So Kepler's First Law is an extremely good approximation.

Kepler's laws are further distorted by that fact that every body in the Solar System attracts every other body with a force that is proportional to its mass. So, for example, the most massive of the planets, Jupiter, exerts a small but (nowadays) measurable distorting effect on the orbit of Mars. Fortunately it was not enough to prevent Kepler from arriving at his laws of planetary motion.

But the worldview held in the time of Kepler has, over the subsequent four centuries, been comprehensively shattered. Kepler himself would barely recognize the Universe we now find we inhabit. Worse than that, he would be totally dismayed by it. It is worth briefly reviewing the discoveries that have brought about this enormous change.

EIGHTEENTH- AND NINETEENTH-CENTURY ASTRONOMY (1701–1900)

First to go was the idea that we live in a closed Universe, surrounded by a celestial sphere. Kepler wished merely to displace the center of the Universe from the Earth to the Sun. The Earth, he was certain, was still virtually at the center of the Universe because the distance to the celestial sphere was so huge. He had no doubt at all that "the whole Universe is enclosed by a spherical shape."[9]

In this respect, as in certain others, his views turned out to be mistaken. Once it was accepted that it was the Earth rather than the celestial sphere that rotated on its axis once every twenty-four hours, the celestial sphere lost its main raison d'être. And the discovery that a number of stars slowly (over decades or centuries) changed their positions relative to other stars provided clear evidence that a celestial sphere did not exist. Gradually, people came to accept Giordano Bruno's suggestion that the stars were at varying distances from the Earth, in a Universe that might well be infinite in extent.

In Kepler's time, only the relative distances of the Sun and the planets were known. (So for example, it was known that the average distance from the Sun to Mars was about 1.52 times as great as that from the Sun to the Earth.) But nobody knew the absolute distances. The first serious attempts to obtain an absolute scale for the size of the Solar System were made following observations of the transit of Venus across the face of the Sun in 1761 and 1769. They depended on the fact that different observers in different places on the planet see the transit at different times. An elaborate calculation (first outlined by the astronomer Edmond Halley many years earlier) then enabled the calculation of the first ever fairly accurate distance to the Sun, which we now know to be about 150 million kilometers away.

Until the late eighteenth and early nineteenth centuries, astronomy consisted largely of the study of the Solar System. The stars and the Universe as a whole formed an interesting backcloth but were too remote to be anything more. This was gradually to change. One of the people who helped to bring this about was Sir William Herschel (1738–1822). Herschel was a German musician who came to England at the age of nineteen as a penniless refugee. He began a successful musical career but gradually came to the view that astronomy was more interesting than music.

Herschel was the first person to undertake systematic telescopic observations of the stars, accidentally discovering the planet Uranus (in 1781) in the process. At a stroke, he doubled the known size of the Solar System—Uranus is about twice as far from the Sun as Saturn. (If there was anyone left who still believed in Kepler's five perfect solids as the basis for the design of the Solar System, this extra planet would surely have destroyed that belief.) Herschel's careful star surveys also provided the first solid observational evidence for the view that we lived in a massive grouping of stars shaped rather like a gigantic disc, an idea already suggested by Thomas Wright (1711–1786) and Immanuel Kant (1724–1804).

A big step forward was made in our understanding of the huge distances to even the nearest stars when Friedrich Bessel (1784–1846) announced in 1838 that he had measured the distance to 61 Cygni, a relatively close star. His result also provided final confirmation (if it were needed) that the Earth does indeed move around the Sun,[10] as his method involved the measurement of the parallax of 61 Cygni. The apparent tiny wobble of this star backward and forward over a year (relative to the much more distant stars beyond it), announced for the first time by Bessel, was merely a reflection of the Earth's own motion around the Sun. Bessel's announcement narrowly beat that from Thomas Henderson (1798–1844), who had measured the distance to Alpha Centauri. Friedrich Georg Wilhelm Struve (1793–1864) also measured the distance to Vega at about the same time.

Astronomical distances are so enormous that astronomers use the distance that light travels in a year as their basic measurement of distance. One light-year is a distance of a little less than 10 *trillion* kilometers, and 61 Cygni is about 11 light-years away. The closest star, Proxima Centauri, is about 4 light-years away. Our Milky Way Galaxy (Herschel's gigantic disc) is about 100,000 light-years in diameter.

Early in the nineteenth century, it had been noted that the motion of Uranus in its orbit around the Sun departed very slightly from what would be expected from Kepler's Second Law. Initially, it moved faster than the law predicted, and then more slowly. The most likely explanation was that this was caused by the gravitational influence of another planet beyond the orbit of Uranus. The calculations for the position of this new planet were made independently by John Adams (1819–1892) and Urbain Le Verrier (1811–1877) and led to the discovery of Neptune in 1846.[11]

Thanks to the work of Charles Lyell (1797–1875) and other geologists, it was also in the nineteenth century that scientists gradually came to realize that we live on a planet that is at least hundreds of millions of years old, rather than merely thousands of years. The age of the Earth is now reliably estimated at a little more than 4.5 *billion* years. This can be calculated using the known decay rates of radioactive elements obtained from a variety of sources—meteorites, rocks brought back from the Moon, and the oldest rocks found on Earth. It is a far cry from Kepler's conviction (and that of all his contemporaries) that the Earth was about 6,000 years old.

Arguably the most important astronomical development of the nineteenth century was that of spectroscopy. In the 1840s, the otherwise eminently sensible French philosopher Auguste Comte (1798–1857) unwisely made the statement that there was no means by which we would ever be able to study the chemical composition of the stars. Within a couple of decades, that is precisely what we were able to do.[12]

The development came about largely because of the work of the chemists Robert Bunsen (1811–1899) and Gustav Kirchhoff (1824–1887) in Germany. It was based on the discovery that every element emits or absorbs light at a characteristic wavelength. One classic demonstration of this is in the ubiquitous sodium-based street lighting, whose characteristic yellow glare does so much to block out our view of the night sky. The sodium atom both emits and absorbs light at a wavelength of about 589 nanometers (within the yellow part of the visible spectrum). All other atoms and molecules also emit and absorb light at their own characteristic wavelengths. For the first time, it became possible to demonstrate that the rest of the Universe was made of exactly the same stuff as we find on Earth. By studying the spectrum of starlight, we now know exactly what the stars are made of.

A particularly good example of the power of spectroscopy was provided by the observation of prominences on the Sun. (Prominences are gigantic plumes of luminous ionized gas that erupt from the Sun's surface into the solar atmosphere. They can normally only be seen around the edge of the Sun during solar eclipses, when the Sun's disc is covered by the Moon.) In 1868, the astronomer Sir Norman Lockyer (1836–1920) had discovered a way to obtain spectra of solar prominences without the need to wait for a solar eclipse. Later that year, he became puzzled by a bright-yellow line in the spectrum of the prominences. The line did not correspond to anything emitted by any known element. Lockyer took the bold step of attributing the line to a new element, which he named *helium* (after the ancient Greek word *Helios*, the god of the Sun). Helium was subsequently discovered on Earth, first by the Italian physicist Luigi Palmieri in 1881, and then (more famously) by Sir William Ramsay in 1895.[13]

TWENTIETH- AND TWENTY-FIRST-CENTURY ASTRONOMY AND COSMOLOGY

It was during the twentieth century that astronomers began to appreciate how utterly small and insignificant we are in the Universe. At the beginning of the twentieth century, it was already realized that our Solar System is a tiny and inconsequential part of our Galaxy. (The first evidence for this had come from William Herschel.) But most astronomers had come to the view that our Galaxy (which we now know contains some 100,000,000,000 stars) was all there was in the Universe. There were, it was true, mysterious clouds of stars referred to as spiral nebulae, but most people had concluded that these were structures within our own Galaxy.

In 1920, a famous debate was held on the subject in the United States. Heber Curtis (1872–1942), a noted American astronomer, took the minority view that the spiral nebulae were separate galaxies in their own right. Harlow Shapley (1885–1972), another American astronomer, who had contributed very significantly to our understanding of the size and structure of our Galaxy (and who had demonstrated that the Sun is not, after all, at the center of our Galaxy), argued the majority view that the spiral nebulae were structures within our Galaxy.

They didn't have to wait long before they found out who was right. In 1924, the American astronomer Edwin Hubble (1889–1953) announced that he had measured the distances to a number of Cepheid variable stars that were embedded within these spiral nebulae, using the powerful new telescope on Mount Wilson, in California. (Cepheid variables are pulsating stars, and their periods of pulsation provide a measurement of their absolute levels of brightness. Their apparent brightnesses can be measured directly, so it is then a simple calculation to determine their distances.) Hubble's calculations showed that the spiral nebulae lay far beyond the outermost boundaries of our own Galaxy. The conclusion was clear—the spiral nebulae were galaxies in their own right. At a stroke, Hubble had shown what some had suspected—that the Universe was vastly larger than we had previously imagined.

Although Hubble examined only a handful of galaxies, a combination of larger telescopes, telescopes in space, and vastly improved ways of collecting light from faint objects mean that we now know there are something like 100,000,000,000 galaxies in the observable Universe (which is in turn part of a much larger—or even infinite—Universe). We have become unbelievably tiny in comparison.

THE BIG BANG THEORY

Five years later, in 1929, Hubble was to announce an even more astonishing discovery. It was already known (thanks to the work of Vesto Slipher [1875–1969] as far back as 1912) that many of these spiral nebulae were moving away from us at high speeds. Hubble plotted a graph of the distances he had measured to several galaxies against their speed of recession using Slipher's unique dataset. He found that he obtained a straight-line graph—in other words, the farther a galaxy was from us, the faster it seemed to be moving away from us.

The possibility (within the context of Albert Einstein's general theory of relativity) of a dynamic Universe had already been predicted by the Russian mathematician Alexander Friedmann (1888–1925) in 1922. It had also been predicted independently by the Belgian astronomer and priest Georges Lemaître (1894–1966) in 1927. Lemaître went on to tie his theory in with Slipher's measurements of speeds of recession of galaxies and with the distance measurements that Hubble had already published. From these measurements, he reached the conclusion that the Universe is expanding. He deserves as much credit as Hubble for this result.[14]

For somebody who was something of a self-publicist, Hubble himself was strangely reluctant to accept the enormous cosmological significance of this discovery. But it soon came to be seen as the first piece of observational evidence for what was later to become known as the Big Bang theory: the idea that the Universe began in a small, hot, and dense state several billion years ago and has been expanding ever since. And by measuring the speed of recession of galaxies, we can work backward to calculate the time at which all those galaxies emerged from this incredibly small and dense state. This now gives us an age for our Universe of 13.8 *billion* years, roughly three times the age of the Earth.

Einstein's general theory of relativity had originally been published in 1916 and is a theory about gravity. It has replaced Isaac Newton's theory of gravity, which had held sway for more than two hundred years. One apparently curious feature of the theory is the notion that empty space can still have interesting properties. It can expand, it can contract, and it can curve. So the Big Bang was *not* seen as some sort of explosion into a preexisting space. It was instead seen as space itself expanding and carrying the galaxies with it.

Further evidence for the Big Bang theory came in 1964, when Arno Penzias (1933–) and Robert Wilson (1936–), two engineers working for Bell Labs in the United States, were using a radio telescope operating at microwave wavelengths to communicate with early communications satellites. They found that they were unable to get rid of background noise in the telescope, regardless of which direction in the sky they pointed it in. They were soon in touch with nearby Princeton physicists Robert Dicke (1916–1997) and James Peebles (1935–), who informed them that they had discovered the Cosmic Microwave Background Radiation.

The CMBR is a direct prediction of the Big Bang theory. According to this theory, when the Universe began in a hot, dense state, it contained both matter and very high-energy gamma radiation. This radiation would not have disappeared as the Universe expanded; instead, its wavelength would have been gradually stretched by the expansion and would by now (several billion years later) be at microwave wavelengths. This is what Penzias and Wilson had detected.

The final piece of evidence for the Big Bang was the fact that it predicted that, over a very wide range of initial conditions, roughly 75 percent (by mass) of the ordinary matter in the Universe should have ended up as hydrogen, and the remaining 25 percent should be helium. (Heavier elements were made later inside stars and still constitute less than 2 percent of the total.) This ratio is just what has been found.

However, the original version of the Big Bang theory was plagued with a number of problems, not the least of which was the question of what exactly went bang, and why. Remarkably, the problems were all solved at a stroke by Alan Guth (1947–),[15] who put forward a theory of inflation in 1981. To a cosmologist, inflation has nothing to do with the increases in the prices of goods and services. It is instead a possible consequence of Einstein's general theory of relativity. In contrast to Newton's theory of gravity, in which gravity is always an attractive force, there can be rare circumstances when gravitational forces in general relativity can be repulsive, rather than attractive.

Alan Guth's stroke of genius was to realize that these circumstances could have applied in a tiny bubble of space in the early Universe. When he did the calculations, he found that the repulsive gravity would have been enormous for a very brief period of time (far less than one second). So this tiny bubble would have increased in size at an exponential rate until it was vastly larger than at the beginning. In the process, a huge amount of gravitational potential energy would have been created. Gravitational energy is negative, so, for total energy to be conserved, positive energy would also have been created in the process—this would have been in the form of matter. (Einstein's famous equation $E = mc^2$ simply says that matter and energy are interchangeable.)

So, as Lawrence Krauss (1954–) and other cosmologists have pointed out, one of the most remarkable facts about the Universe is that the total amount of matter (positive energy) plus gravitational energy (negative) in the Universe seems to be equal to zero.[16] The answer to the question "Why is there something rather than nothing?" is that there may well be a very real sense in which there is (overall) nothing in the Universe. As Guth has said, it looks as though the Universe really is the ultimate free lunch. The cosmologist Max Tegmark

(1967–) expressed the same sentiment when he described the Universe as a gigantic Ponzi scheme.[17]

Inflation solves the problem of what went bang, and of where all the matter came from, and lots more besides. It is not yet a complete certainty that inflation occurred, but many cosmologists accept it as an elegant solution to a number of puzzles. Most interesting of all, many theories of inflation have a fascinating consequence. They imply that new universes are coming into existence all the time and will continue to do so forever into the future. Collectively, we refer to all these universes (perhaps infinite in number) as the Multiverse.

There is no reason to suppose that the vast majority of these other universes in the Multiverse will have the same set of physical constants as those that exist in our own Universe. The physical constants will vary from one universe to another. Our Universe will be one of the tiny minority (possibly, it must be said, infinite in number) where the relative values of the physical constants are such as to allow the existence of chemical complexity, which seems to be a fundamental requirement for life, and hence for intelligent life. Almost all the other universes are unlikely to be able to sustain complexity, and are therefore likely to be without life.

If a Multiverse theory is correct, the origin of our Universe no longer seems to require a divine creator. Given an infinite amount of time and an infinite space, our highly unlikely Universe was bound to come into existence eventually. The balance of probabilities has changed. We have moved on from a position, five hundred years ago, where it seemed beyond question that the Universe had been deliberately created, and that humans had been placed at its center. It now seems more likely that we are an unimaginably tiny part of a Universe that arose entirely as a result of random processes, rather than conscious design. It may take a while for humanity to get used to this new—and probably inevitable—perspective.

So it now seems that, until recently, we were in a similar position in our ideas about the Universe to the one that Kepler held about the Solar System. Kepler believed that the distances in the Solar System had been divinely ordained, and that it ought to be possible to calculate them from first principles. We now know that these distances are essentially random—the numerous other solar systems that we have now discovered beyond our own show no pattern, and serve to confirm this.

In the same way, almost all scientists until a short time ago thought that, if only we could come up with a single unified "theory of every-thing," it should be possible to calculate the various physical constants of the Universe from first principles. If, however, the theory of the Multiverse is correct, then the physical constants that happen to exist in our Universe are entirely random—just like the orbits of planets. Other universes will have entirely different sets of physical constants. We may have been making exactly the same mistake as Kepler, but on a far vaster scale.

CONCLUSION: THE KEPLER MISSION

On March 6, 2009, at 10:49 p.m. EST, four hundred years after Kepler published his first two laws of planetary motion, NASA's Kepler Mission was launched. It was to carry out one of the most fascinating scientific studies in recent years. Its main aim was to discover Earth-size planets in orbit around stars in what has come to be known as the Goldilocks zone. In other words, it was looking for planets that were neither too hot nor too cold, but that (like Goldilocks's porridge) were just right. These would be rocky planets where water would be found in liquid form and where life might therefore also be present.

It carried out its search by pointing continuously at the same area of the sky (in the region of the constellations of Cygnus and Lyra), and looking at a small sample of about 150,000 individual stars. As seen from Earth, the plane in which planets orbit their stars will be aligned at entirely random angles with respect to a line between Earth and the star. Only for about 1 percent of all the stars that have planetary systems will those systems (by chance) happen to orbit in a plane that lies along (or very close to) that Earth-star line. Planets going around those stars will seem to move across the surface of those stars once every complete orbit. In doing so, they will cause a very slight dip in the amount of light we receive from the star. These dips are what the

Kepler Mission has been detecting. We can be confident that a dip is caused by a planetary transit only if the dip is detected (at least) three times and the time period between each pair of dips is the same.

We know the period of revolution of the planet (its year) from the observations of the time between dips in the star's light. Then (as we know the mass of the star) we can calculate the distance of the planet from the star, from Kepler's Third Law of planetary motion. So (as we also know the surface temperature of the star) we can then estimate whether the planet lies within the Goldilocks zone. We can also estimate the size of the planet by looking at the size and shape of the dip in brightness. From gravitational interactions with the star, we can work out the planet's mass, and hence its density, so we know whether it is rocky (and Earth-like) or gaseous.

Unfortunately, the Kepler Mission came to an end in May 2013, when a key component in the satellite failed. But sufficient results had already been gathered to show that planets are commonplace in our Galaxy. And although an agreed-upon figure is yet to be established, estimates for the number of Earth-like planets in the Goldilocks zone in our Galaxy are now in the billions. We also know that there are something like 100,000,000,000 galaxies in the observable Universe. So this means that—at a conservative estimate—there are 100,000,000,000,000,000,000,000 Earth-like planets on which life could have arisen in the observable Universe. Our planet is certainly not exceptional. Ironically, Kepler himself would have been horrified by this conclusion, as he always believed that the Solar System had a unique status in the Universe.[18]

Primitive life emerged on Earth about 3.8 billion years ago, and intelligent life has only emerged relatively recently. It may be that simple unicellular life is common throughout the Universe, but that the emergence of more complex life-forms (and therefore of intelligent life) is a relatively rare event. Nevertheless, even if only one planet in every hundred galaxies in the observable Universe will at some stage have been home to intelligent life, that still implies the existence of 1,000,000,000 intelligent civilizations in the last few (or coming few) billions of years. Disappointingly, even if they evolved at about the same time as us (which is statistically unlikely), all of them will—almost certainly—be so far away that it is difficult to see how we could ever make contact with them.

These civilizations will no doubt be very different from ours, in ways we cannot even begin to imagine. But of one thing we can be sure—if they have developed science, they, too, will all independently have discovered Johannes Kepler's three laws of planetary motion.

APPENDIX

SUMMARY OF KEPLER'S TRAVELS

December 27, 1571: Birth in Weil der Stadt.

Late 1575: Family move to Leonberg.

1578: Move to Ellmendingen.

1579: Move back to Leonberg.

October 1584: Boarder at Adelberg seminary.

November 1586: Boarder at Maulbronn seminary.

October 1589: Student at Tübingen University.

March 1594: Leaves Tübingen.

April 1594: Arrives in Graz to become a schoolmaster.

Early 1596: Returns to Württemberg (to visit grandfathers and discuss book).

July 1596: Returns to Graz.

September 1598: One-month exile from Graz.

January 1600: Leaves Graz for Prague (meeting with Tycho Brahe).

June 1600: Returns to Graz (via Vienna).

September 1600: Leaves Graz for Prague with wife and stepdaughter.

April 1601: Returns to Graz (to settle father-in-law's affairs).

August 1601: Returns to Prague (Tycho dies October 1601).

April 1612: Leaves Prague for Linz.

October 1617: Travels to Württemberg (following his mother).

December 1617: Returns to Linz.

1620–21: Linz/Württemberg (witch trial).

November 1626: Departure from Linz with family. Family found lodgings in Regensburg. Kepler arrives in Ulm.

November 1627: Returns to family in Regensburg.

December 1627: Visits Prague to see Ferdinand II.

May 1628: Leaves Prague for Regensburg (then Linz briefly).

July 1628: Arrives in Sagan.

October 1630: Leaves Sagan.

November 1630: Arrival in Regensburg, and death on November 15.

ACKNOWLEDGMENTS

I am grateful to a number of publishers and authors for allowing me to use quotations from their books, as follows:

William H. Donahue: Translations of *Astronomia Nova* and *Optics.*

Dover: *Kepler*, by Max Caspar.

Philosophical Library: *Johannes Kepler: Life and Letters*, by Carola Baumgardt.

University of Chicago Press: Translation of *The Sidereal Messenger*, by Albert Van Helden.

Oxford University Press: Translation of Kepler's "Preface to the *Rudolphine Tables*," by Owen Gingerich and William Walderman.

Cambridge University Press: Translation of *A Defence of Tycho against Ursus*, by N. Jardine.

and to the following for allowing me to reproduce diagrams, as follows:

Euratlas: Three maps of parts of Europe as they were in 1600.

Bayerische Akademie of Science: Kepler's model of the Solar System, his diagram of Great Conjunctions, his eclipse viewing device, his table of world history, and his frontispiece to the *Rudolphine Tables*.

Syndics of Cambridge University Library: An illustration from the *Nuremberg Chronicle*.

SkyMap Pro software: Two star maps.

I am also very grateful to Prometheus Books for having informed me that no permissions were required for the various quotations I have used from some of their books.

All quotations are set either in quotation marks or as indented text and are referenced with endnotes. All diagrams are similarly referenced.

All photographs, unless otherwise noted, were taken by me, and I own the copyrights.

To the best of my knowledge, I have obtained all necessary permissions for the contents of this book. If there are any errors or omissions, Prometheus Books will be pleased to insert the appropriate acknowledgment in any subsequent printing of this book.

AUTHOR'S NOTE ON THE SPELLING OF "RUDOLPH"

T he Holy Roman Emperor Rudolph II played a significant role in Kepler's life. In Kepler's time, and for centuries afterward, the spellings both of "Rudolph" and of the *Rudolphine Tables* named after him used "ph." In recent decades, however, many authors have chosen to spell his name as "Rudolf." In this book, I have decided to use the spelling that Kepler himself would have recognized. For the same reason, I am using "Alphonso" rather than "Alfonso."

ABOUT THE SOURCES

There are two main sources of primary historical documents on Kepler. They are: *Omnia Opera*, published by Heyder & Zimmer (Frankfurt) and edited by Christian Frisch; and *Gesammelte Werke*, published by C. H. Beck (Munich) and originally edited by Walther von Dyck and Max Caspar. Both sources were issued in several volumes over periods of several years or decades.

Omnia Opera (1858–1871) is the earlier work and is not particularly well arranged, but it has the merit of being easily available on the Internet. *Gesammelte Werke* (1937–2012) is more comprehensive and much better structured, but it is expensive and difficult to get hold of in Britain and has only slowly been appearing on the Internet. Most of the text in both sources is in Latin. A small amount is in medieval German. Most of Kepler's key works have now been translated into English.

All sources are listed in the bibliography. However, the most important primary source documents and translations (mostly of works contained within the above volumes) are as follows:

Carola Baumgardt, *Johannes Kepler: Life and Letters* (New York: Philosophical Library, 1951).

Max Caspar, *Kepler* (New York: Dover, 1993).

Nicolaus Copernicus, *On the Revolutions of Heavenly Spheres*, trans. Charles Glenn Wallis (Amherst, NY: Prometheus Books, 1995).

J. L. E. Dreyer, *A History of Astronomy from Thales to Kepler* (Mineola, NY: Dover, 1953).

J. L. E. Dreyer, *Tycho Brahe: A Picture of Scientific Life & Work in the Sixteenth Century* (Edinburgh: Black, 1890).

Galileo, *The Sidereal Messenger*, trans. Albert Van Helden (Chicago: University of Chicago Press, 1989).

Nicholas Jardine, *The Birth of History and Philosophy of Science* (Cambridge: Cambridge University Press, 2009). The book contains Jardine's translation of Kepler's *A Defence of Tycho against Ursus*.

Kepler, *Astronomia Nova (The New Astronomy)*, trans. William H. Donahue (Cambridge: Cambridge University Press, 1992).

Kepler, *Epitome of Copernican Astronomy*, trans. Charles Glenn Wallis (Amherst, NY: Prometheus Books, 1995).

Kepler, *The Harmony of the World*, trans. Charles Glenn Wallis (Amherst, NY: Prometheus Books, 1995).

Kepler, *Optics*, trans. William H. Donahue (Santa Fe, NM: Green Lion Press, 2000).

Kepler, "Preface to the *Rudolphine Tables*." Translated by Owen Gingerich and William Walderman (London: *Quarterly Journal of the Royal Astronomical Society* 13, 1972).

Anyone wishing to find out more about Galileo, Kepler's fellow Copernican, is recommended to read David Wootton's *Galileo: Watcher of the Skies* (New Haven, CT, and London: Yale University Press, 2010).

NOTES

INTRODUCTION

There are numerous books covering the history of astronomy up to the time of Kepler. Perhaps the most detailed and thorough is still J. L. E. Dreyer's *A History of Astronomy from Thales to Kepler* (Mineola, NY: Dover, 1953). This and other relevant books for this period are included in the bibliography. Specific notes are as follows:

1. Bertrand Russell, *History of Western Philosophy* (London: Unwin, 1969), p. 49.

2. Because of this lack of contemporary records, some have even doubted whether he actually existed.

3. There is a fascinating discussion about Pythagoras on "Pythagoras," *In Our Time*, BBC Radio 4, accessed December 14, 2014, http://www.bbc.co.uk/programmes/b00p693b.

4. Carola Baumgardt, *Johannes Kepler: Life and Letters* (New York: Philosophical Library, 1951), p. 46.

5. Ibid., p. 41.

6. J. L. E. Dreyer, *A History of Astronomy from Thales to Kepler* (Mineola, NY: Dover, 1953), pp. 174–76. The achievement is also noted in Michael Hoskin, ed., *The Cambridge Concise History of Astronomy* (Cambridge: Cambridge University Press, 1999), p. 26, and in Paul Murdin and Margaret Penston, eds., *The Canopus Encyclopedia of Astronomy* (Bristol, UK: Canopus, 2004), p. 145.

7. But not everybody. The influential third-century Christian theologian Lactantius (mentioned by Copernicus in his *On the Revolutions of Heavenly Spheres*, p. 7, and by Kepler in his *Astronomia Nova*, p. 66) fiercely defended the view that the Earth was flat.

8. It is probably the existence of these seven wandering bodies in the heavens that have, via the Babylonians, given us our seven-day week.

9. Dreyer goes into this in considerable detail in *A History of Astronomy from Thales to Kepler*, pp. 53–86 and 108–22. The topic is also covered in Hoskin, *The Cambridge Concise History of Astronomy*, pp. 27–29, and in Arthur Berry, *A Short History of Astronomy* (Mineola, NY: Dover, 1961), pp. 26–29, and is briefly and perfectly summarized in Murdin and Penston, *The Canopus Encyclopedia of Astronomy*, pp. 18, 332–33.

10. Hoskin, *The Cambridge Concise History of Astronomy*, pp. 34–36.

11. Ibid., pp. 39–46.

12. The role of the Arabs and Persians in astronomy during medieval times is very well covered in ibid., pp. 50–62.

13. Referred to, for example, in Dreyer, *A History of Astronomy from Thales to Kepler*, p. 273. The precise wording of the quotation is not known.

14. A copy of the book, open at this page, is on display at the museum in Tübingen Schloss.

15. Dreyer, *A History of Astronomy from Thales to Kepler*, p. 305.

16. The word *nepotism* comes from the Italian *nipote*, meaning "nephew" (or "grandson"), because popes of the time often appointed nephews to positions of power. Pope Calixtus III appointed as a cardinal his nephew Rodrigo Borgia, who went on to become Pope Alexander VI between 1492 and 1503.

17. Dreyer, *A History of Astronomy from Thales to Kepler*, p. 138. This early suggestion is also noted in Hoskin, *The Cambridge Concise History of Astronomy*, p. 34.

18. This was apparent to Copernicus even in his earlier "Commentariolus."

19. Strictly, the Copernican system replaces the epicycles of the outer planets and the deferents of the inner planets with the motion of the Earth. But the huge simplification is still the case.

20. Nicolaus Copernicus, *On the Revolutions of Heavenly Spheres*, trans. Charles Glenn Wallis. (Amherst, NY: Prometheus Books, 1995), p. 26.

21. The epicyclet, a mini-epicycle that perpetuated the idea of circles on circles.

22. This is a literal reading of Copernicus's statement in the dedication of the book to Pope Paul III, that his work had been ready for almost four periods of nine years. Copernicus, *On the Revolutions of Heavenly Spheres*, p. 5.

23. His "Commentariolus."

24. Nicholas Jardine, *The Birth of History and Philosophy of Science* (Cambridge: Cambridge University Press, 2009), p. 152.

25. Recorded, for example, in Dreyer, *A History of Astronomy from Thales to Kepler*, p. 319.

26. Copernicus, *On the Revolutions of Heavenly Spheres*, pp. 25–26.

27. Andrew White, *History of the Warfare of Science with Theology in Christendom* (Amherst, NY: Prometheus Books, 1993), p. 126. A reference to this also appears in Thomas Kuhn, *The Copernican Revolution* (Cambridge, MA: Harvard University Press, 1957), p. 191; in Dreyer, *A History of Astronomy from Thales to Kepler*, p. 352–53; and in Angus Armitage, *The World of Copernicus* (Dublin: Mentor Books, 1956), p. 94.

28. The latest archaeological evidence indicates that Joshua's brutal invasion probably never took place—see Israel Finkelstein and Neil Asher Silberman, *The Bible Unearthed: Archaeology's New Vision of Ancient Israel and the Origin of Its Sacred Texts* (New York: Touchstone, 2002).

29. White, *History of the Warfare of Science with Theology in Christendom*, p. 127. A reference to this also appears in Kuhn, *The Copernican Revolution*, p. 191; in Dreyer,

A History of Astronomy from Thales to Kepler, pp. 352–53; and in Armitage, *The World of Copernicus*, p. 94.

30. There is a plaque dedicated to Melanchthon on the wall of the Burse at Tübingen University. It gives his date of birth and the dates when he studied and taught at the university.

31. Biblical passages that indicate that the Earth is fixed and the Sun moves include Ps. 19:5—"The Sun ... rejoices like a strong man to run his race"; Ps. 93:1—"Thou has fixed the Earth immovable and firm"; Ps. 96:10—"He has fixed the Earth, firm, immovable"; Ps. 104:5—"Thou didst fix the Earth on its foundation, so that it can never be shaken"; 1 Chron. 16:30—"He has fixed the Earth firm, immovable"; Josh. 10:13—"So the Sun stood still and made no haste to set for almost a whole day"; Matt. 5:35—"nor by Earth, for it is [God's] footstool." A footstool could hardly be racing through space around the Sun.

32. The apocryphal nature of the quote is convincingly demonstrated by Edward Rosen, in his usual scholarly way, in *Copernicus and His Successors* (London: Hambledon 1995), p. 161 onward.

33. Copernicus, *On the Revolutions of Heavenly Spheres*, p. 4.

34. It can be shown that the motion of a planet on an eccentric (an offset circle) is mathematically equivalent to the motion of that planet on a suitable epicycle, the center of which orbits the Earth, even though these models would require different physical mechanisms.

35. Kepler refers to this in at least two places: (a) *A Defence of Tycho against Ursus* in Jardine, *The Birth of History and Philosophy of Science*, p. 150: "the author of the preface was Andreas Osiander"); and (b) his *Astronomia Nova*, original title page, p. 28.

36. Kepler certainly thought so. In 1601, he wrote that "Osiander's plan has clearly succeeded for the past sixty years" (Jardine, *The Birth of History and Philosophy of Science*, p. 152).

37. I am grateful to *Wikipedia* for this much clearer version of Tycho's original illustration. *Wikipedia*, s.v. "Tycho Brahe," last modified June 22, 2015, accessed July 15, 2015, https://en.wikipedia.org/wiki/Tycho_Brahe.

38. Possibly indirectly named after Copernicus, who was thought of as a Prussian, rather than a Pole. See Johannes Kepler, "Preface to the *Rudolphine Tables*," trans. Owen Gingerich and William Walderman, *Quarterly Journal of the Royal Astronomical Society* 13 (1972), p. 366. Dreyer considers that they were so named after Reinhold's patron, Duke Albrecht of Prussia (Dreyer, *A History of Astronomy from Thales to Kepler*, p. 345).

39. The *Alphonsine Tables* were far from perfect, as Kepler pointed out: "The reliability of the tables was less than their reputation." Kepler, *Rudolphine Tables*, p. 365.

40. Again, as pointed out by Kepler. Ibid., p. 366.

41. On the road leading up to Tübingen Schloss can be found the house where Maestlin lived, and in the Schloss Museum is a painting of Maestlin.

42. Dr. Stephen Pumfrey, senior lecturer in history at Lancaster University, has made a detailed study of William Gilbert and has concluded that Gilbert's writings, notably those in Book II of *De Mundo*, make sense only on the assumption that he was a closet Copernican.

43. There were some other indications of acceptance of Copernicus's ideas. Robert Recorde, a Welsh mathematician, wrote a book called *The Castle of Knowledge*, published in 1556, in which he has a student say to his master—about Copernicanism—that he "desire[s] not to hear such vain fantasies, so far against common reason & repugnant to the consent of all the learned multitude of writers." The master replies that the student is "too young to be a good judge in so great a matter" and that "you were best to condemn no thing that you do not well understand" and that another time "I will so declare his supposition that you shall not only wonder to hear it, but also peradventure be as earnest then to credit it as you are now to condemn it." So Robert Recorde can probably be counted in as another closet Copernican. (There is a reference to Recorde in Dreyer, *A History of Astronomy from Thales to Kepler*, p. 346.)

CHAPTER 1: KEPLER'S EARLY LIFE

1. Unless otherwise stated, biographical information and quotations in chapter 1 come from one of the following two sources:

(a) Johannes Kepler, *Omnia Opera*, ed. Christian Frisch, 8 vols. (Frankfurt: Heyder & Zimmer, 1858–1871), vol. 8, pt. 2, p. 668 onward; and (b) Johannes Kepler, *Gesammelte Werke*, ed. Walther von Dyck and Max Caspar, 22 vols. (Munich: C. H. Beck, 1937–2012), 14:275–77. Letter to Fabricius dated October 1, 1602. Both translated by Mario Miccichè.

2. From Kepler, *Gesammelte Werke*, 19:320. Times were measured from noon, rather than midnight, so 02.30 is 2.30 p.m. For the same reason, the time of conception Kepler gives as 16.37 is actually 4.37 a.m. the following morning. I am very grateful to Professor Owen Gingerich for confirming that this is the case.

3. The documents give May 15, because the wedding was in the morning. Using the modern convention, this becomes May 16.

4. Kepler gives the period of the pregnancy as 224 days and 10 hours (strictly, it should be 9 hours, 53 minutes if it is to line up precisely with the times he gives for his conception and birth). By subtraction, this gives the date of conception as May 17 (using our modern convention of starting each new day at midnight). Again, Owen Gingerich has kindly confirmed this. Arthur Koestler, on the first page of *The Watershed: A Biography of Johannes Kepler* (London: Heinemann, 1961), wrongly gives this date as May 16, because he is following the old convention of measuring dates from midday.

5. Owen Chadwick, *Penguin History of the Church—The Reformation* (London: Penguin, 1990), p. 143.

6. Information from Herr Wolfgang Schütz at the Kepler Museum in Weil der Stadt. The source of the information is the records in Stuttgart, as most of Weil's records were destroyed in the fire of 1648. If anything, they are overestimates, as both sides would have wanted to exaggerate their numbers. However, Caspar's and Koestler's figure of 200 citizens and their families (Max Caspar, *Kepler* [New York: Dover, 1993], p. 32, and Koestler, *The Watershed*, p. 18) is almost certainly an underestimate.

7. Caspar, *Kepler*, p. 35. The brothers are also shown in the family tree to be found at the Kepler Museum in Weil der Stadt.

8. Information from the Kepler Museum in Weil der Stadt.

9. Kepler makes a number of references to his poor eyesight. The problem of multiple vision is referred to in his *Optics*, trans. William H. Donahue. (Santa Fe, NM: Green Lion Press, 2000), pp . 216–17, where Kepler says: "Hence it is that those who labor under this defect see a doubled or tripled object in place of a single narrow and far distant one. Hence in place of a single moon, ten or more are presented to me."

10. Tycho's observations of the 1577 comet and his conclusions on its significance are discussed in J. L. E. Dreyer's *A History of Astronomy from Thales to Kepler* (Mineola, NY: Dover, 1953), pp. 365–66, and in Michael Hoskin, ed., *The Cambridge Concise History of Astronomy* (Cambridge: Cambridge University Press, 1999), p. 102.

11. Kepler, *Optics*, pp. 284–85: "The cause, however, is entirely in the refractions, so that the redness is nothing else but the illumination of the Moon by the Sun's rays, transmitted through the density of the air, and refracted inwards towards the axis of the shadow."

12. Information from the Leonberg Museum, which was formerly Kepler's Latin school. The plaque on the museum's wall proclaims the dates when Kepler was at the school.

13. Information from the Kepler Museum in Weil der Stadt.

14. Information for this paragraph is from the Maulbronn Monastery guidebook.

15. Kepler, *Gesammelte Werke*, 19:328: "Homo iste hoc fato natus est, ut plerumque rebus difficilibus tempus terat, a quibus alij abhorrent, etc." Self-Characterization (hereafter this title is used to indicate the source of Kepler's writings about himself).

16. Ibid., pp. 328–29: "Mathemata prae caeteris studijs amavit. . . . Horologium coeleste confinxit." Self-Characterization.

17. Ibid., p. 329: "In Theologia statim initio de praedestinatione incepit et in Lutheri sententiam de servo arbitrio incidit. . . . Etiam gentibus non omnimodam damnationem propositam existimavit, motus speculation misericordiae divinae." Self-Characterization.

18. Ibid., pp. 336–37, "doluit sibi ob jam admissam vitae impuritatem negatum esse prophetiae honorem . . . Erat autem quarundam contionum recitatio." Self-Characterization.

19. Ibid., p. 329: "Tenax in re pecuniaria nimium, in oeconomia rigidus, minutissimorum censor, quibus omnibus tempus extrahitur." Self-Characterization.

20. Ibid., p. 336: "Habet homo iste naturam undiquaque caninam, etc." Self-Characterization.

21. From Kepler's autobiographical notes in *Omnia Opera*, vol. 8, pt. 2, p. 671. "1589: . . . my father . . . finally went into exile to die."

22. Johannes Kepler, *Astronomia Nova*, trans. William H. Donahue (Cambridge: Cambridge University Press, 1992), p. 183: "And since I was supported at the expense of the Duke of Württemberg."

23. Caspar, *Kepler*, p. 43.

24. The official Tübingen guidebook contains much useful background on the University.

25. These quotations are from *Omnia Opera*, vol. 8, pt. 2, p. 676: "amore saucius laboravi" and "mihi oblatum conjugium virginis."

26. Caspar, *Kepler*, p. 43.

27. Ibid., p. 44.

28. Chadwick, *Penguin History of the Church—The Reformation*, pp. 144–45.

29. This is certainly the view taken by Max Caspar: *Kepler*, p. 46. Edward Rosen takes a different view: *Copernicus and His Successors* (London: Hambledon Press, 1995), p. 218, although he is—for once—not very convincing.

30. Caspar, *Kepler*, p. 47.

31. Kepler, *Gesammelte Werke*, 1:9: "quo labore me facilè liberare potuisset Joachimus Rheticus."

32. Kepler, *Gesammelte Werke*, 19:3.

33. Ibid., p. 3: "etliche Universitäten."

34. Johannes Kepler, *Epitome of Copernican Astronomy*, trans. Charles Glenn Wallis (Amherst, NY: Prometheus Books, 1995), p. 10. Evidence has recently emerged that the young Kepler, while still at Tübingen, did sign the Formula of Concord. In light of all we know of Kepler, this seems totally out of character. We have to assume that it must have been under duress.

35. Kepler, *Gesammelte Werke*, 19:337. "De religion cupide in vulgus disserit."

36. Caspar, *Kepler*, p. 49.

37. Arthur Beer and Peter Beer, *Proceedings of Conferences Held in Honor of Kepler* (Oxford: Pergamon, 1975), p. 28.

38. This is certainly also the view of Alexandre Koyré, *The Astronomical Revolution* (London: Routledge, 2009), p. 379; J. L. E. Dreyer, *Tycho Brahe: A Picture of Scientific Life & Work in the Sixteenth Century* (Edinburgh: Black 1890), p. 289; and Arthur Koestler, *The Watershed: A Biography of Johannes Kepler*, p. 33. Much later in life (Johannes Kepler, *Selbstzeugnisse*, trans. Franz Hammer [Stuttgart: Verlag, 1971], p. 63.) Kepler once claimed that he had kept his theological views to himself, but this contradicts what he himself had said earlier.

39. Kepler, *Astronomia Nova*, chap. 7, p. 184.

40. Ibid., p. 183.

41. Ibid., pp. 183–84.

42. Ibid.: "and saw my comrades, whom the Prince, upon request, was sending to foreign countries, stalling in various ways out of love for their country I—being hardier—quite maturely agreed with myself that wherever I was destined I would promptly go."

CHAPTER 2: GRAZ (1594–1600)

1. Paul Murdin and Margaret Penston, eds., *The Canopus Encyclopedia of Astronomy* (Bristol, UK: Canopus, 2004), p. 63.

2. Owen Chadwick, *Penguin History of the Church—The Reformation* (London: Penguin, 1990), p. 302.

3. Johannes Kepler, *Omnia Opera*, ed. Christian Frisch, 8 vols. (Frankfurt: Heyder & Zimmer, 1858–1871), vol. 8, pt. 2, p. 677.

4. Ibid. Kepler says here that his expenses were paid on April 21; it is reasonable to assume that he arrived in Graz immediately before that and that the date he gives is in terms of the new Gregorian calendar, which he supported. Subsequent dates from this source and from his autobiographical letter to David Fabricius are Gregorian, rather than Julian. Max Caspar, in his biography of Kepler, converts the April 21 (Gregorian) date back to April 11 (Julian). I am grateful to Professor Owen Gingerich for confirming this.

5. Ibid.

6. Ibid.

7. Max Caspar, *Kepler* (New York: Dover, 1993), p. 56.

8. Usually attributed to the late and great John Wheeler.

9. Johannes Kepler, *Gesammelte Werke*, ed. Walther von Dyck and Max Caspar, 22 vols. (Munich: C. H. Beck, 1937–2012), 19:332: "Ex eo taediosa aut certè perplexa et minus intelligibilis efficitur ejus oratio." Self-Characterization.

10. Kepler, *Gesammelte Werke*, 8:20: "gravis."

11. Caspar, *Kepler*, p. 60. See also Kepler's letter to Maestlin: Kepler, *Gesammelte Werke*, 13: 19.

12. Peter Machamer, ed., *The Cambridge Companion to Galileo* (Cambridge: Cambridge University Press, 1998), p. 19.

13. Caspar, *Kepler*, p. 61. See also Kepler, *Gesammelte Werke*, 1:9.

14. The time between the successive lining up of two planets, as seen by an observer on the Sun, is called the synodic period (S). It can be calculated from $(1/S) = (1/T_1) - (1/T_2)$, where T_1 and T_2 are the orbital periods of the two planets. If you substitute $T_1 = 11.9$ years and $T_2 = 29.5$ years, then S = 20 years.

15. Kepler, *Gesammelte Werke*, 1:11. "Igitur die 9. vel 19. Iulij anni 1595." Kepler seems a little uncertain about the exact date. In a letter to his friend Fabricius, he gives the date as July 17, rather than July 19.

16. Ibid., p. 10.

17. Ibid., p. 12. Reprinted with permission.

18. Caspar, *Kepler*, p. 62. "I believe it was by divine ordinance that I obtained by chance that which previously I could not reach by any pains."

19. Ibid., p. 63.

20. Johannes Kepler, *Astronomia Nova*, trans. William H. Donahue (Cambridge: Cambridge University Press, 1992), chap. 7, p. 184: "So from that time I began to think seriously of comparing observations. In 1597, I wrote to Tycho Brahe, asking his opinion of my little book."

21. Kepler, *Gesammelte Werke*, 1:48: "sed quod in tanta distantia nemo miretur."

22. Ibid., p. 29.

23. Ibid., p. 10, "Inter Iouem et Martem interposui nouum Planetam."

24. Murdin and Penston, *The Canopus Encyclopedia of Astronomy*, p. 78.

25. Kepler, *Gesammelte Werke*, 1:26–27. Reprinted with permission.

26. Ibid., p. 4.

27. Ibid., p. 4: "Hic te Pythagoras docet omnia quinque figuris."

28. Ibid., p. 16. "Amat illa simplicitatem, amat vnitatem. Nunquam in ipsa quicquam ociosum aut superfluum extitit."

29. Ibid., p. 16: "grauissima praeceptoris mei MAESTLINI clarissimi Mathematici authoritate."

30. Ibid., p. 55.

31. Ibid., p. 78: "Annus igitur Christi 1595 si referatur in 5572."

32. Whether Kepler picked up this idea from *Narratio Prima* or some other source, or whether he thought of it independently is an interesting question. However, even if the former is the case, Kepler is certainly the first person who pursued the idea energetically.

33. Kepler, *Gesammelte Werke*, 1:70.

34. Ibid., p. 201.

35. Ibid., p. 76: "Ergo in medietate, etc."

36. Ibid., 13:86.

37. Carola Baumgardt, *Johannes Kepler, Life and Letters* (New York: Philosophical Library, 1951), p. 37.

38. Kepler, *Gesammelte Werke*, 8:20: "post duos menses reuersus sum in Styriam," i.e., in July.

39. Ibid., p. 20.

40. Ibid., p. 20: "et errant qui absurditate moti dogmatis Copernicani conatibus meis intercederent."

41. Kepler, *Omnia Opera*, 1:37: "Pro egregia phantasia et erudito invento idem Hafenreffer agnoscit, sed sanctae scripturae et ipsi veritati contrariari omnino et simpliciter putat."

42. Caspar, *Kepler*, p. 68.

43. Kepler, *Gesammelte Werke*, 8:20: "nam pro Keplero expresserunt Replereum."

44. Ibid., p. 9.

45. James R. Voelkel, *The Composition of Kepler's* Astronomia Nova (Princeton, NJ: Princeton University Press, 2001) gives the example of two contemporary thinkers, Johannes Praetorius and Helisaeus Roeslin, who did not agree at all with Kepler's ideas. He shows that even those whom Kepler quoted as agreeing with the book were in reality (and with the exception of Maestlin and Georg Limnaeus) not doing so.

46. Kepler, *Gesammelte Werke*, 14:275–77: October 1, 1602, letter to Fabricius.

47. Ibid., vol. 8, p. 20: "GALILAEVS Patauio," i.e., Galileo of Padua.

48. Baumgardt, *Johannes Kepler, Life and Letters*, pp. 38–39.

49. Ibid., pp. 40–41.

50. In *The Birth of History and Philosophy of Science* (Cambridge: Cambridge University Press, 2009), pp. 10–11, Nicholas Jardine points out that Ursus's cosmology differs from Tycho's in three significant respects: Ursus accepts the daily rotation of the Earth, he places the orbit of Mars outside that of the Sun (so that the two do not intersect), and he suggests that the stars may be at different distances. So even if Ursus got his basic idea from Tycho, it is clear from Jardine's comments that he did not simply copy it blindly.

51. There are numerous accounts of the life of Tycho Brahe. Perhaps the best and most detailed of these is J. L. E. Dreyer, *Tycho Brahe: A Picture of Scientific Life and Work in the Sixteenth Century* (Edinburgh: Black, 1890).

52. Dreyer, *Tycho Brahe*, pp. 217–22 and pp. 235–36.

53. See n. 73 for dates of letters.

54. Kepler, *Gesammelte Werke*, 13:197–201: "Placet is sane non mediocriter, et ingenii tui acumen sagaxque studium non obscure hinc elucent. ... Ingeniosa certe et succincta est speculatio, planetarum distantias et circuitus symmetriis regularium corporum, uti facis adstringere, et plurima in his satis consentire videntur, non obstante."

55. Jardine, *The Birth of History and Philosophy of Science*, p. 14.

56. Ibid., p. 12.

57. Ibid., p. 17.

58. Kepler, *Omnia Opera*, vol. 8, pt. 2, p. 677.

59. Kepler, *Gesammelte Werke*, 14:275–77: October 1, 1602, letter to Fabricius.

60. Caspar, *Kepler*, p. 72.

61. Self-Characterization, Kepler, *Gesammelte Werke*, 19:328–37.

62. Frau Boockmann, a member of the Kepler Commission in Munich, has kindly provided me, via the Weil Museum staff, with a definitive list of names and dates of all Kepler's children.

63. Owen Chadwick, *Penguin History of the Church—The Reformation* (London: Penguin, 1990), p. 310.

64. Kepler, *Gesammelte Werke*, 8:20: "Quanquam Eclipsis magna Solis in Dodecatemorio Piscium, etc."

65. Chadwick, *Penguin History of the Church—The Reformation*, p. 310.

66. One of the frustrations of the historian is in not necessarily being able to establish definitive reasons for particular events. Caspar, in *Kepler* (pp. 80–81), outlines all these very reasonable possibilities and concludes that it was probably the respect that prominent Catholics had developed for him that enabled Kepler to be allowed back to Graz. But we cannot be certain.

67. See n. 62 above.

68. Caspar, *Kepler*, p. 84, and Chadwick, *Penguin History of the Church—The Reformation*, p. 310.

69. Kepler, *Gesammelte Werke*, 8:21: "ipse salarium," etc.

70. Baumgardt, *Johannes Kepler, Life and Letters*, p. 54.

71. Kepler, *Gesammelte Werke*, 13:292 and Kepler, *Omnia Opera*, 1:50, as translated in Caspar, *Kepler*, p. 87.

72. Kepler, *Gesammelte Werke*, 8:21: "loco mihi oblato in comitatu suo."

73. The sequence of letters between Kepler and Tycho at this stage can be a little confusing. It was as follows:

K to T: Dec. 13, 1597 (received Mar. 1598).

T to K: Apr. 11, 1598 (received Feb. 1599—the copy to Maestlin was received by him Aug. 1598).

K to T: Feb. 19, 1599.

T to K: Dec. 9, 1599 (arrived after K had already left Graz).

T to K: Jan. 26, 1600 (on hearing that K was now in Prague).

Kepler refers to several more letters but—if they ever existed—they have not survived.

74.

Johannes Kepler (1571–1630) taught here at the former protestant apprentice school 1594–1599 as a teacher of mathematics. In Graz he wrote his first astronomical work "The Mystery of the Universe" which made him famous throughout the western world. In 1600 in the wake of the Counter-Reformation he had to leave Graz, and became the employee and successor of Tycho Brahe in Prague at the court of the Emperor Rudolph II. The evangelical apprentice school was closed and converted into a convent for Poor Clares nuns.

(Translated by Rachel Love.)

CHAPTER 3: KEPLER AND TYCHO

1. Part of Benatky Castle is now a small astronomy museum. Included in the exhibits is a wooden carving of Tycho.

2. Max Caspar, *Kepler* (New York: Dover, 1993), p. 100. The letter is in Johannes Kepler, *Gesammelte Werke*, ed. Walther von Dyck and Max Caspar, 22 vols. (Munich: C. H. Beck, 1937–2012), 14:107 onward.

3. Johannes Kepler, "Preface to the *Rudolphine Tables*," trans. Owen Gingerich and William Walderman, *Quarterly Journal of the Royal Astronomical Society* 13 (1972): 368: "I came to Tycho at Benatky Castle . . . only in February of 1600."

4. Carola Baumgardt, *Johannes Kepler: Life and Letters* (New York: Philosophical Library, 1951), p. 61.

5. Kepler, *Gesammelte Werke*, 19:37–41.

6. Johannes Kepler, *Optics*, trans. William H. Donahue (Santa Fe, NM: Green Lion Press, 2000), p. 171: "my friend."

7. Arthur Koestler, *The Watershed: A Biography of Johannes Kepler* (London: Heinemann, 1960), p. 116. Kepler's groveling letter is in Kepler, *Gesammelte Werke*, 14:114 onward.

8. Caspar, *Kepler*, p. 107.

9. Kepler, "*Rudolphine Tables*," p. 369.

10. Caspar, *Kepler*, p. 107.

11. Kepler, "*Rudolphine Tables*," p. 368.

12. Baumgardt, *Johannes Kepler: Life and Letters*, p. 61. Letter to Herwart von Hohenburg.

13. Caspar, *Kepler*, p. 102. Letter to Herwart von Hohenburg.

14. Kepler, *Optics*, p. 347.

15. Replicas of this device can be found both at the Bundesrealgymnasium in Graz and at the Kepler Museum in Regensburg.

16. Peter H. Wilson, *Europe's Tragedy* (London: Penguin, 2010), p. 72.

17. Kepler, *Gesammelte Werke*, 8:21: "Receptis autem breui aliquot BRAHEI epistolis, etc." Tycho even signed his letter of August 28, 1600, "tui amantissimus," best translated as "ever devoted": Kepler, *Gesammelte Werke*, 14:148.

18. Baumgardt, *Johannes Kepler: Life and Letters*, p. 56. Letter of September 19, 1600.

19. 3:34. Also in Kepler, "*Rudolphine Tables*," p. 368. Letter to Longomontanus of 1605.

20. Baumgardt, *Johannes Kepler: Life and Letters*, p. 62 (letter of December 16, 1600).

21. Kepler, *Omnia Opera*, 1:56: "Ad haec Maestlinus non respondit . . . usque ad annum 1605."

22. Nicholas Jardine, *The Birth of History and Philosophy of Science* (Cambridge:

Cambridge University Press, 2009), p. 135: "I also said it to Ursus himself, to his face, as soon as I came to Prague last January. However, I concealed my name.... And having finally revealed my name, I parted from him peaceably."

23. Ibid., p. 147: "Ursus holds hypotheses to be only fictitious."

24. Ibid., p. 148: "he maintains those he himself throws up to be supported by the infallible authority of the Scriptures."

25. Ibid., p. 135: "Every hypothesis whatsoever, if we examine it minutely, yields some consequence which is entirely its own and is not shared with any other hypothesis."

26. Ibid., p. 216: "In Copernicus' work, a most beautiful regularity is revealed in all these things, the cause must likewise be contained therein."

27. Ibid., pp. 206–207.

28. Kepler, *Omnia Opera*, 3:34; also in Kepler, *Gesammelte Werke*, 15:134–43. Letter to Longomontanus of 1605.

29. Kepler, *"Rudolphine Tables,"* p. 368: "I had returned to Graz on account of my legacy."

30. Norman Davies, *Europe: A History* (London: Pimlico, 1997), p. 530.

31. "Rudolph II," *In Our Time*, BBC Radio 4, January 31, 2008, accessed January 6, 2015, http://www.bbc.co.uk/programmes/b008tsj9.

32. Ibid.

33. Davies, *Europe: A History*, p. 529.

34. R. J. W. Evans, *Rudolph II and His World* (London: Thames and Hudson, 1997), pp. 84–115.

35. Kepler, *Gesammelte Werke*, 8:21: "quibus BRAHEVS à RVDOLPHO nomen esse voluit."

36. Johannes Kepler, *Astronomia Nova*, trans. William H. Donahue (Cambridge: Cambridge University Press, 1992), p. 157: "and on his deathbed asked me, whom he knew to be of the Copernican persuasion, that I demonstrate everything in his hypotheses."

37. This paragraph is based on Kepler's own account of Tycho's death, as written up in the volume of observations for 1600 and 1601. It is reproduced in J. L. E. Dreyer, *Tycho Brahe: A Picture of Scientific Life and Work in the Sixteenth Century* (Edinburgh: Black, 1890), pp. 386–87.

38. Kepler, *Gesammelte Werke*, 14:275–77: October 1, 1602, letter to Fabricius.

39. Joshua Gilder and Anne-Lee Gilder put forward the view that Kepler poisoned Tycho in *Heavenly Intrigue: Johannes Kepler, Tycho Brahe, and the Murder Behind One of History's Greatest Scientific Discoveries*, (London: Doubleday, 2004).

40. Caspar, *Kepler*, p. 102.

41. "Astronomer Tycho Brahe 'Not Poisoned,' Says Expert," BBC News, accessed January 6, 2015, http://www.bbc.co.uk/news/science-environment-20344201.

CHAPTER 4: PRAGUE (1600–1612)

1. Carola Baumgardt, *Johannes Kepler: Life and Letters* (New York: Philosophical Library, 1951), p. 67. Letter to Maestlin.

2. Baumgardt, *Johannes Kepler: Life and Letters*, p. 95. Letter to Fabricius of October 1, 1602. Kepler confesses to his illegitimate borrowing of Tycho's observations in Johannes Kepler, *Gesammelte Werke*, ed. Walther von Dyck and Max Caspar, 22 vols. (Munich: C. H. Beck, 1937–2012), 15:232: "Non diffiteor, me Tychone mortuo, haeredibus vel absentibus, vel parum peritis, observationum relictarum tutelam mihi confidenter, et forsan arroganter usurpasse" ("I confess that I on Tycho's death, his heirs being either absent or not sufficiently experienced, bequeathed the observations to myself for their secure protection, and perhaps arrogantly usurped them").

3. Baumgardt, *Johannes Kepler: Life and Letters*, pp. 95–96. Letter to Fabricius of February 7, 1604.

4. Johannes Kepler, *Astronomia Nova*, trans. William H. Donahue (Cambridge: Cambridge University Press, 1992), pp. 43–44: "Especially his liberty in disagreeing with Brahe in physical arguments."

5. Arthur Koestler, *The Watershed: A Biography of Johannes Kepler* (London: Heinemann, 1961), p. 162.

6. Kepler, *Astronomia Nova*, chap. 6, p. 157. Kepler makes *Astronomia Nova* far more complicated than it needs to be by describing the motions of the planets in terms of Tycho's system and Ptolemy's system, as well as his own.

7. Ibid. Note on p. 51 of introduction to Kepler's *Astronomia Nova*.

8. Bertrand Russell, *History of Western Philosophy* (London: Unwin, 1969), p. 516.

9. William H. Donahue, *Selections from Astronomia Nova* (Santa Fe, NM: Green Lion Press, 2008).

10. Bruce Stephenson, *Kepler's Physical Astronomy* (Princeton, NJ: Princeton University Press, 1987).

11. James R. Voelkel, *The Composition of Kepler's Astronomia Nova* (Princeton, NJ: Princeton University Press, 2001).

12. Kepler, *Astronomia Nova*, p. 78.

13. Baumgardt, *Johannes Kepler: Life and Letters*, p. 61.

14. Max Caspar, *Kepler* (New York: Dover, 1993), p. 126.

15. The equation of a circle is $x^2/r^2 + y^2/r^2 = 1$, where x and y are the coordinates of any point on its circumference, and r is its radius. The equation of an ellipse is $x^2/a^2 + y^2/b^2 = 1$, where x and y are as before; a is half the distance between the two farthest points on the ellipse, and b is half the distance between the two closest points on the ellipse. So a circle is simply the special case of an ellipse where a = b.

16. Kepler, *Astronomia Nova*, p. 185.

17. The eight-stage analysis is this author's own way of explaining Kepler's

work. There are numerous summary accounts in the literature of how he achieved his results. One of the best is in J. L. E. Dreyer, *A History of Astronomy from Thales to Kepler* (New York: Dover, 1953). Others include A. J. Aiton, "The Elliptical Orbit and the Area Law," in Arthur Beer and Peter Beer, *Proceedings of Conferences Held in Honor of Kepler* (Oxford: Pergamon, 1975); Donahue, *Selections from Astronomia Nova*; Stephenson, *Kepler's Physical Astronomy*; Koestler, *The Watershed*.

18. If the Earth moved around the Sun at a uniform angular rate, then the times between an equinox and a solstice would all be equal. But the following table (which takes into account the fact that each equinox and solstice takes place at an instant in time, not over a whole day) shows that they are not quite the same:

	Days
September 23, 2010, to December 21, 2010	89.85
December 21, 2010, to March 20, 2011	88.99
March 20, 2011, to June 21, 2011	92.75
June 21, 2011, to September 23, 2011	93.66
One year	365.25

19. In fact, the position of the mean Sun in Copernicus's system corresponded to the center of the Earth's orbit, but there was no reason to assume that this empty point in space should have any physical significance.

20. Kepler, *Gesammelte Werke*, 1:76: "computauit non à centro Solis, sed à centro orbis Magni." Although this may simply have been because his five perfect solids fit better that way!

21. Kepler, *Astronomia Nova*, chap. 7, pp. 184, 185: "But when I found out during the first week that, like Ptolemy and Copernicus, he made use of the Sun's mean motion, while the apparent motion would be more in accord with my little book [i.e., *Mysterium Cosmographicum*], I begged the master to allow me to make use of the observations in my own manner." And on this issue, "At the beginning there was great controversy between us."

22. Ibid., p. 48: "Now my first step in investigating the physical causes of the motions was to demonstrate that [the planes of] all the eccentrics intersect in no other place than the very center of the solar body (not some nearby point), contrary to what Copernicus and Brahe thought."

23. Owen Gingerich, *The Eye of Heaven* (New York: American Institute of Physics, 1993), p. 339. J. L. E. Dreyer had already made essentially the same suggestion in *A History of Astronomy from Thales to Kepler*, p. 410.

24. Strictly, Mars orbits once every 686.98 days, not exactly 687 days.

25. Kepler, *Astronomia Nova*, p. 85.

26. Four observations were needed, rather than three, because Kepler initially

did not make the assumption of Copernicus—and in effect Ptolemy as well—that the Sun and the equant point were equidistant from the center. If equidistance were assumed, then only three observations would have been needed. However, Kepler's method at that stage showed that the two distances were not equal.

27. At March 6, 1587, June 8, 1591, August 25, 1593, and October 31, 1595: Kepler, *Astronomia Nova*, p. 257. He describes his basic assumptions as follows:

> In imitation of the ancients, physical causes aside, it is posited that the course of the planet is a circle, that within this circle there exists some point about which the planet traverses equal angles in equal times, and that between this point and the center of the Sun lies the center of the planetary circle, at some unknown distance. With these assumptions, four acronychal [i.e., opposition] observations with zodiacal positions and their time intervals are used to seek out, by a most laborious method, the zodiacal positions of those centers, their distances from the center of the Sun, and the ratios of the two eccentricities, both to one another and to the radius of the circle.

Kepler, *Astronomia Nova*, p. 85. Kepler actually demonstrated his calculations using all three systems (Ptolemaic, Tychonic, and Copernican), which can for certain purposes be regarded as geometrically identical.

28. Ibid., p. 256.

29. For the years 1580, 1582, 1585, 1589, 1597, 1600, 1602, and 1604. Kepler observed these last two oppositions himself—Tycho was by then dead.

30. Kepler, *Astronomia Nova*, p. 276: "You see then, O studious reader, that the hypothesis."

31. Ibid., p. 281.

32. The two observations in latitude were at the oppositions of January 30, 1585, and August 25, 1593 (Kepler, *Astronomia Nova*, p. 282), and they allowed Kepler to calculate the eccentric anomaly for Mars by an independent method. The oppositions were about 8.5 Earth years and 4.5 Mars years apart, so they took place on almost opposite sides of the Sun. This helped in the geometry of Kepler's calculations. The alternative method then gave different results!

33. Kepler says that Ptolemy's level of accuracy was a full 10 minutes. Kepler, *Astronomia Nova*, p. 286: "Now Ptolemy professed not to go below 10 minutes."

34. Ibid., p. 284.

35. Ibid., p. 286.

36. For example, in Kepler's *Mysterium Cosmographicum*, he states that neither Ptolemy nor Copernicus used an equant point for the Earth.

37. Baumgardt, *Johannes Kepler: Life and Letters*, p. 10. Introduction by Albert Einstein.

38. The dates Kepler chose were March 5, 1590, January 21, 1592, December

8, 1593, and October 26, 1595, all 687 days apart. He deliberately chose these dates because on all four occasions, Mars was very close to the ecliptic, so nobody could argue that variations in latitude were muddling the calculation. Kepler, *Astronomia Nova*, p. 316.

39. Kepler, *Astronomia Nova*, p. 51.

40. Donahue, *Selections from Astronomia Nova*, p. 51.

41. Kepler points this out as early as chapter 4. Kepler, *Astronomia Nova*, p. 135: "Therefore the point of the equant is nothing but a geometrical short cut for computing the equations from an hypothesis that is clearly physical."

42. Ibid., p. 380.

43. Caspar, *Kepler*, p. 138. From letter to David Fabricius of October 11, 1605.

44. Kepler, *Astronomia Nova*, p. 402.

45. Ibid., p. 55.

46. Ibid., p. 57.

47. Ibid., p. 56.

48. Baumgardt, *Johannes Kepler: Life and Letters*, p. 43.

49. Kepler, *Astronomia Nova*, p. 67. Kepler did not recognize the term Solar System, but this seems to be his meaning.

50. Ibid., pp. 386–87: "The body of the Sun must move . . . upon its axis."

51. Ibid., p. 388: "The planetary globes [are] . . . inclined to rest or to the privation of motion."

52. For example, in Johannes Kepler, *The Harmony of the World*, trans. Charles Glenn Wallis (Amherst, NY: Prometheus Books, 1995), p. 244, he states, "The unvarying rotation of the Sun . . . (concerning which the sunspots are evidence)." But it would be wrong to give the impression that he consistently guessed correctly. For example, he wrongly deduced that the Sun must be the densest of all the bodies in the Solar System (Kepler, *Astronomia Nova*, p. 390). In fact, it is a lot less dense than the four rocky planets—Mercury, Venus, Earth, and Mars—and of comparable density to the four gas giants—Jupiter, Saturn, Uranus, and Neptune.

53. Ibid., p. 405.

54. Ibid., p. 412.

55. Johannes Kepler, *Optics*, trans. William H. Donahue (Santa Fe, NM: Green Lion Press, 2000), pp. 347–53.

56. Kepler, *Astronomia Nova*, p. 375: "At other points, there appears a very small discrepancy."

57. Ibid., pp. 417–18.

58. Ibid., p. 419.

59. Ibid., pp. 419–20.

60. For example, in ibid., p. 120: "very nearly circular," or ibid., p. 407: "Let the orbit of the planet be a circle, as has been believed until now."

61. Ibid., p. 453.

62. Letter of July 4, 1603, to Fabricius. Kepler, *Gesammelte Werke*, 14:409–35.

63. Kepler, *Astronomia Nova*, chap. 58, p. 575.

64. Donahue, *Selections from Astronomia Nova*, p. 96. In his translation of Kepler's *Optics*, Donahue also points out that the very first use of the word *focus* is in this book (i.e., *Optics*).

65. As mentioned in the introduction, the *Prutenic Tables* were the tables widely used at the time for calculating future planetary positions. They had been compiled by Erasmus Reinhold a few years after the death of Copernicus. Although the tables were based on Copernicus's theory of a Sun-centered Universe, Reinhold treated the idea— just like Andreas Osiander—not as something to be regarded as true but merely to provide a more reliable basis for computation than the much earlier *Alphonsine Tables*.

66. Kepler, *Astronomia Nova*, p. 48.

67. Mark 3:28–30; Matt. 12:30–32; Luke 12:8–10.

68. See the biblical references in n. 31 to the introduction of this book.

69. Kepler, *Astronomia Nova*, p. 59.

70. Ibid. Copernicus uses exactly the same quotation (from Virgil's *Aeneid*) in his *On the Revolutions of Heavenly Spheres*, book 1, chap. 8, p. 17.

71. One example that Kepler quotes is Psalm 24. The author of this psalm undoubtedly believed that a flat Earth was positioned firmly on top of a huge ocean. But Kepler dismisses this idea on the (incorrect) grounds that this passage was never meant literally.

72. Kepler valiantly attempts to reinterpret Psalm 104, which states that "thou didst fix the Earth on its foundation, so that it can never be shaken." He does so somewhat implausibly, by stating that it is a commentary on the six days of creation described in chapter 1 of the book of Genesis. (In his time, the Genesis creation story was still taken literally.)

73. Kepler, *Astronomia Nova*, p. 53.

74. Ibid., p. 30, Dedication.

75. Ibid., pp. 140–41.

76. Ibid., p. 34, Dedication.

77. Ibid.

CHAPTER 5: PRAGUE—MANY NEW THINGS

1. Johannes Kepler, *Gesammelte Werke*, ed. Walther von Dyck and Max Caspar, 22 vols. (Munich: C. H. Beck, 1937–2012), 19:331. Self-Characterization: "Quod multa incipit nova prioribus imperfectis." Changed from third-person singular to first-person singular in the translation.

2. Johannes Kepler, *Astronomia Nova*, trans. William H. Donahue (Cambridge: Cambridge University Press, 1992), p. 256.

3. Johannes Kepler, *Optics*, trans. William H. Donahue (Santa Fe, NM: Green Lion Press, 2000), p. 22. Proposition 9 in a long list of propositions. Kepler's typically long-winded way of expressing the inverse square law was as follows: "The ratio that holds between spherical surfaces, a larger to a smaller, in which the source of light is [at the] center, is the same as the ratio of strength or density of the rays of light in the smaller to that in the more spacious spherical surface: that is, inversely."

4. Ibid., p. 179.

5. Ibid., p. 191: "a crystalline or aqueous globe ... next to a glazed window ... a white piece of paper behind the globe, distant ... by a semi-diameter of the globe ... the window [is] depicted with perfect clarity upon the paper but in an inverted position."

6. Ibid., p. 224. Kepler approvingly quotes Porta, who says that "the likeness is sent in through the pupil as through the opening of a window."

7. Ibid., p. 216.

8. Ibid., p. 217.

9. Ibid., pp. 217–18.

10. Ibid., p. 216.

11. Ibid., p. 260: "This is evidence that some parts of it are low, others more raised up."

12. Ibid., p. 266: "We therefore say that the Earth, by its gleaming light, sent to it from the Sun, casts its rays on the ... night in the lunar body no less than ... (in exactly the same way) the full Moon illuminates our nights in Earth."

13. Ibid., pp. 284–85.

14. Kepler, *Gesammelte Werke*, 13:184: "Ego sum Lutheranus astrologus." Letter to Maestlin of March 15, 1598.

15. Ibid., 4:146. From Kepler's *Tertius Interveniens*, published in 1610. Title page: "nicht das Kindt mit dem Badt außschütten."

16. Carola Baumgardt, *Johannes Kepler: Life and Letters* (New York: Philosophical Library, 1951), p. 51.

17. Ibid., pp. 51–52.

18. Ibid., p. 52.

19. J. V. Field, "A Lutheran Astrologer: Johannes Kepler," *Archive for History of Exact Sciences* 31 (London, 1984): 189–272.

20. Johannes Kepler, "Preface to the *Rudolphine Tables*," trans. Owen Gingerich and William Walderman, *Quarterly Journal of the Royal Astronomical Society* 13 (1972): 367.

21. The three cosmologists to win the Nobel Prize for this groundbreaking discovery were Saul Perlmutter, Brian P. Schmidt, and Adam G. Riess. A brief account of their discovery, the background to it, and its huge significance is given at the Nobel Prize website, accessed June 29, 2015, http://www.nobelprize.org/nobel_prizes/physics/laureates/2011/press.html.

22. See, for example, Michael Hoskin, ed., *The Cambridge Concise History of Astronomy* (Cambridge: Cambridge University Press, 1999), p. 96.

23. Johannes Kepler, *L'Étoile Nouvelle dans le Pied du Serpentaire (On the New Star)*, trans. Jean Peyroux (Paris: A. Blanchard, 1998), pp. 55–57, 197–98.

24. Kepler, *Gesammelte Werke*, 1:194.

25. Kepler, *L'Étoile Nouvelle dans le Pied du Serpentaire*, p. 76.

26. Johannes Kepler, *Omnia Opera*, ed. Christian Frisch, 8 vols. (Frankfurt: Heyder & Zimmer, 1858–1871), 2:607–608. This was in a covering letter, rather than the book itself.

27. Modern scholarship confirms this date for Herod's death. See, for example, Geza Vermes, *The Nativity* (London: Penguin, 2006), p. 92.

28. Kepler, *Omnia Opera*, 4:177: "Stellam igitur, quae Magos perduxit ad Christi praesepe; utpote biennio antiquiorem navitate Christi, hac circumstantia nostrae huic stellae fuisse comparandam." Author's translation: "The star, therefore, which led the wise men to the manger of Christ two years after the birth of Christ, this was comparable to our star."

29. See, for example, Vermes, *The Nativity*, pp. 93–96; or Bart Ehrman, *The New Testament* (Oxford: Oxford University Press, 2008), p. 127; or E. P. Sanders, *The Historical Figure of Jesus* (London: Penguin, 1995), pp. 85–87. The two biblical accounts of Jesus's birth (Matthew and Luke) are mutually contradictory in several respects, but this is only widely known among New Testament scholars. For understandable reasons, the Christian Church is unwilling to draw these problems to the attention of its members.

30. There are a number of translations of *Somnium* into English. These include Johannes Kepler, *Kepler's Dream*, trans. John Lear and Patricia Kirkwood (Berkeley and Los Angeles: University of California Press, 1965); and Johannes Kepler, *Kepler's Somnium*, trans. Edward Rosen (New York: Dover, 1967). Another version, for Kindle, can be downloaded from Amazon.

31. Kepler, *Omnia Opera*, vol. 8, pt. 1, p. 23.

32. Kepler, *Gesammelte Werke*, 4:297: "ut in ipsius Wackherii gratiam, etiam Astronomiam novam."

33. For example, Kepler, *Somnium: Kepler's Dream*, p. 91, n. 8.

34. Kepler, *Gesammelte Werke*, 4:270.

35. Baumgardt, *Johannes Kepler: Life and Letters*, p. 75.

36. Ibid., p. 78: "My opinion that there are unseen comets in the sky is disputed by many."

37. See, for example, Hoskin, The Cambridge Concise History of Astronomy, pp. 147–48.

38. This has been extensively demonstrated in James R. Voelkel, *The Composition of Kepler's* Astronomia Nova (Princeton, NJ: Princeton University Press, 2001), chap. 8.

39. See, for example, Paul Murdin and Margaret Penston, eds., *The Canopus Encyclopedia of Astronomy* (Bristol, UK: Canopus, 2004), pp. 60–61.

40. In Kepler, *Gesammelte Werke*, 4:304 (his *Dissertatio Cum Nuncio Sidereo*), Kepler describes Bruno's ideas as *"horridae"*—horrible.

41. Baumgardt, *Johannes Kepler: Life and Letters*, p. 78.

42. Max Caspar, *Kepler* (New York: Dover, 1993), p. 157.

CHAPTER 6: 1610—THE YEAR OF THE TELESCOPE

1. See, for example, Paul Murdin and Margaret Penston, eds., *The Canopus Encyclopedia of Astronomy* (Bristol, UK: Canopus, 2004), p. 231.

2. There is a very good biography of Harriot by John W. Shirley, *Thomas Harriot: A Biography* (Oxford: Clarendon, 1983).

3. Galileo, *The Sidereal Messenger*, trans. Albert Van Helden (Chicago: University of Chicago Press, 1989), p. 40.

4. J. L. E. Dreyer, *A History of Astronomy from Thales to Kepler* (New York: Dover, 1953), p. 57.

5. Galileo, *The Sidereal Messenger*, p. 62.

6. Ibid., p. 58.

7. Ibid., p. 57.

8. Ibid., p. 64.

9. Ibid., p. 31.

10. Johannes Kepler, *Gesammelte Werke*, ed. Walther von Dyck and Max Caspar, 22 vols. (Munich: C. H. Beck, 1937–2012), 4:288: "Matthaeus Wackherius a Wakhenfelsz, de curru mihi ante habitationem meam nunciasset. Tanta me incessit admiratio, absurdissimi acroamatis consideratione, tanti orti animorum motus (quippe ex inopinato decisa antiqua inter nos liticula) ut ille gaudio, ego rubore, risu uterque ob novitatem confusi, ille narrando ego audiendo vix sufficeremus." Author's translation: Matthaeus Wackherius of Wackenfels announced [this] from his carriage in front of my dwelling. So great was my surprise at this piece of news, ... as he with joy, I with shame, both with laughter because of the novelty of the situation, that I was hardly able to listen to the narrative."

11. Ibid., 4:289: "Wackherio contrà visum, haud dubiè circa fixarum aliquas circumire novos hos planetas (quale quid iam à multo tempore mihi ex Cardinalis Cusani et Giordani Bruno speculationibus objecerat)."

12. Ibid., 4:290.

13. Ibid., 4:291.

14. See, for example, Murdin and Penston, *The Canopus Encyclopedia of Astronomy*, p. 320. Phobos and Deimos are named after the Greek gods Horror and Terror, appropriate deities to accompany the god of war.

15. Kepler, *Gesammelte Werke*, 4:296.

16. Ibid., 4:297.

17. See, for example, Murdin and Penston, *The Canopus Encyclopedia of Astronomy*, p. 275.

18. Johannes Kepler, *Optics*, trans. William H. Donahue (Santa Fe, NM: Green Lion Press, 2000), pp. 262–63.

19. Kepler, *Gesammelte Werke*, 4:299: "Nam profectò consentaneum est, si sunt in Luna viventes creaturae."

20. Ibid., 4:299: "Potuit te huius aeris Lunaris admonere."

21. Ibid., 4:306: "sed proculdubio Iovialibus creaturis."

22. Ibid., 4:307: "etiam jovem circa suum volvi axem."

23. Ibid., 4:304: "horridae."

24. Ibid., 4:303: "Satis igitur hinc clarum est, Corpus huius nostri Solis inaestimabili mensura esse lucidius, quàm universas fixas."

25. Ibid., 4:302: "esse praecipuum mundi sinum."

26. Johannes Kepler, *Optics*, trans. William H. Donahue (Santa Fe, NM: Green Lion Press, 2000), p. 22.

27. Kepler, *Gesammelte Werke*, 4:308: "Sint illi infiniti mundi dissimiles nostri."

28. Ibid., 4:305: "Da naves, aut vela caelesti aurae accommoda, erunt qui ne ab illa quidem vastitate sibi metuant."

29. Carola Baumgardt, *Johannes Kepler: Life and Letters* (New York: Philosophical Library, 1951), p. 86.

30. Johannes Kepler, *Omnia Opera*, ed. Christian Frisch, 8 vols. (Frankfurt: Heyder & Zimmer, 1858–1871), 2:462.

31. The illustration is taken from Galileo, *The Sidereal Messenger*, trans. Albert Van Helden. University of Chicago Press assures me that no copyright adheres to the picture and that I can reproduce it, but I should nevertheless like to acknowledge its source and to compliment the author on having produced such an admirably clear diagram.

32. Kepler was aware of Galileo's argument and repeated it in his *Epitome of Copernican Astronomy*, trans. Charles Glenn Wallis (Amherst, NY: Prometheus Books, 1995), p. 68.

33. Kepler, *Omnia Opera*, 2:465.

34. Kepler's guesses are listed in *Omnia Opera*, 2:468.

35. See, for example, Murdin and Penston, *The Canopus Encyclopedia of Astronomy*, p. 184.

36. David Wootton, *Galileo: Watcher of the Skies* (New Haven, CT, and London: Yale University Press, 2010), pp. 125–31. In this book, Wootton has produced a very well-researched and thorough biography of Galileo.

37. Kepler, *Omnia Opera*, 2:773.

38. Max Caspar, *Kepler* (New York: Dover, 1993), p. 197.

39. John Donne, *The Complete Poetry and Selected Prose of John Donne*, ed. Charles M. Coffin (London: Modern Library [Random House], 2001. "Ignatius His Conclave" is at 56 percent of the Kindle edition.

CHAPTER 7: LINZ (1612-1626)

1. Carola Baumgardt, *Johannes Kepler: Life and Letters* (New York: Philosophical Library, 1951), p. 94.

2. Max Caspar, *Kepler* (New York: Dover, 1993), pp. 174–75.

3. Johannes Kepler, *Gesammelte Werke*, ed. Walther von Dyck and Max Caspar, 22 vols. (Munich: C. H. Beck, 1937–2012), vol. 16. Kepler's letters to the duke dated March 19, 1611, trans. Martin Ziegler and Dennis Emerson.

4. Ibid., vol. 16. Kepler's letters to the duke's mother dated March 19, 1611, trans. Martin Ziegler and Dennis Emerson.

5. Baumgardt, *Johannes Kepler: Life and Letters*, p. 101.

6. Caspar, *Kepler*, p. 206.

7. Baumgardt, *Johannes Kepler: Life and Letters*, pp. 102–104.

8. Peter H. Wilson, *Europe's Tragedy* (London: Penguin, 2010), p. 56.

9. Caspar, *Kepler*, p. 211.

10. Ibid., p. 212.

11. Ibid., p. 188.

12. Owen Chadwick, *Penguin History of the Church—The Reformation* (London: Penguin, 1990), pp. 144–45. The puzzle mentioned in n. 34 of chap. 1, that the young Kepler had actually signed the Formula of Concord just before leaving Tübingen, becomes even more of a puzzle at this stage. Why did nobody apparently remind Kepler that he had in fact signed this document?

13. Caspar, *Kepler*, p. 219: "non una via vel vice de incolumitate (safety) mea periclitatus (danger) fui."

14. Ibid., p. 261.

15. Ibid., p. 226.

16. Baumgardt, *Johannes Kepler: Life and Letters*, p. 114.

17. The information on Kepler's children, from which the family tree has been constructed, was kindly supplied by Frau Boockmann, a member of the Kepler Commission in Munich, via Frau Claudia Hehmann of the Kepler Museum in Weil der Stadt. The year of Anna Maria's death is not known but was probably after 1638.

18. Caspar, *Kepler*, p. 233.

19. The quotation comes from Celestino Bianchi et al., *Le Opere di Galileo Galilei* (Florence, Italy: Barbera, 1972), vol. 19, p. 320.

20. Kepler first heard about this from his friend Quietanus Remus, a physician in Emperor Matthias's court. Baumgardt, *Johannes Kepler: Life and Letters*, p. 141.

21. Caspar, *Kepler*, p. 298.

22. Baumgardt, *Johannes Kepler: Life and Letters*, pp. 145–46.

23. This is usually translated as *The Harmony of the World*, but *The Harmony of the Universe* seems more apposite. There is also an arcane dispute about whether the Latin word *Harmonices* is intended to be in the genitive singular ("of the harmony") or in the nominative plural ("harmonies"). I have assumed the former.

24. Baumgardt, *Johannes Kepler: Life and Letters*, pp. 147–48.

25. Kepler expresses this most clearly in Johannes Kepler, *Epitome of Copernican Astronomy*, trans. Charles Glenn Wallis (Amherst, NY: Prometheus Books, 1995), p. 74: "The apparent movement of the Sun has 365 days, which is the mean measure between Venus's period of 225 days and Mars's period of 687 days. Therefore does not the nature of things shout out loud that the circuit in which those 365 days are taken up has the mean position between the circuits of Mars and of Venus around the Sun; and thus this is not the circuit of the Sun around the Earth . . . but the circuit of the Earth around the resting Sun." He had previously used very similar wording in *Astronomia Nova* (p. 53) to make the same point.

26. Johannes Kepler, *The Harmony of the World*, trans. Charles Glenn Wallis (Amherst, NY: Prometheus Books, 1995), p. 411.

27. Ibid., p. 406: "However, it is not definitely [i.e., completely] equal, as I once dared to promise for eventually perfected astronomy."

28. Ibid., p. 440.

29. Ibid., p. 391.

30. Laplace, for example, writes: "Il est affligeant pour l'esprit humain de voir ce grand homme, même dans ses derniers ouvrages, se complaire avec délices dans ses chimériques spéculations, et les regarder comme l'âme et la vie de l'astronomie."

31. Kepler, *Epitome of Copernican Astronomy*, p. 5.

32. Ibid.

33. Ibid., pp. 139–43.

34. Ibid., p. 78. The author has constructed the following table using both Kepler's figures and modern values:

	Kepler's values			Modern values (setting Europa distance = 5)		
	Distance	Time	D^3 / T^2	Distance	Time	D^3 / T^2
Io	3	1d 18.5hrs	8.6	3.14	1.77d	9.9
Europa	5	3d 13.3hrs	9.9	5	3.55d	9.9
Ganymede	8	7d 2hrs	10.2	7.97	7.15d	9.9
Callisto	14	16d 18hrs	9.8	14.03	16.7d	9.9

35. Ibid., p. 10.

36. Johannes Kepler, "A Defence of Tycho against Ursus," trans. Nicholas Jardine, in *The Birth of History and Philosophy of Science* (Cambridge: Cambridge University Press, 2009), p. 146.

37. Kepler, *Gesammelte Werke*, 8:9: "denique quicquid fere librorum. . . ."

38. Ibid., p. 21: "Ita omnis mihi vitae. . . ."

39. Ex. 22:18. A similar sentiment is expressed in Lev. 20:27.

40. "Witchcraft," *In Our Time*, BBC Radio 4, October 21, 2004, accessed January 6, 2015, http://www.bbc.co.uk/programmes/p004y2b0.

41. Caspar, *Kepler*, p. 241.

42. Johannes Kepler, *Omnia Opera*, ed. Christian Frisch, 8 vols. (Frankfurt: Heyder & Zimmer, 1858–1871), vol. 8, pt. 1, p. 46: "Si verum est, inquam, quod de sagis tradunt pleraque tribunalia, quod illae transportentur per aerem, erit forte et hoc possibile, ut corpus aliquod Terris divulsum importetur in Lunam." Author's translation: "If it is true, I say, as tribunals for the most part say of witches, that they can be transported through the air, it will be, perhaps, possible to make a body separated from earth be conveyed to the Moon."

43. Johannes Kepler, *Optics,* trans. William H. Donahue (Santa Fe, NM: Green Lion Press, 2000), p. 150: "Mount Hoeberg, near Horba, is notorious for . . . gatherings of witches."

44. Caspar, *Kepler*, pp. 241–43.

45. Ibid., p. 244.

46. Ibid., pp. 244–47.

47. Ibid., pp. 174–75.

48. Ibid., pp. 247–53.

49. Kepler, *Omnia Opera*, vol. 8, pt. 2: "Totum annum in causa matris insumsi."

50. Baumgardt, *Johannes Kepler: Life and Letters*, p. 161.

51. Caspar, *Kepler*, pp. 253–56.

52. Translated from the German by Rachel Love. The German version of the poem can be found in Wolfgang Schütz, *Kepler und die Nachwelt* (Weil der Stadt, Germany: Scharpf, 2009), p. 6. This book also contains information about the various portraits of Kepler, and Wolfgang Schütz has kindly supplemented this in personal communications with the author.

53. Norman Davies, *Europe: A History* (London: Pimlico, 1997), p. 568.

54. This was actually the Second Defenestration. The first had taken place nearly 200 years earlier, in 1419.

55. Peter H. Wilson, *Europe's Tragedy: A New History of the Thirty Years War* (London: Penguin, 2010), p. 273.

56. Davies, *Europe: A History*, p. 563.

57. Arthur Beer and Peter Beer, *Proceedings of Conferences Held in Honor of Kepler* (Oxford: Pergamon, 1975), p. 153.

58. Baumgardt, *Johannes Kepler: Life and Letters*, p. 139.

59. Caspar, *Kepler*, p. 309.

60. Ibid., p. 311.

61. Baumgardt, *Johannes Kepler: Life and Letters*, p. 154.

62. Ibid., pp. 151–52

63. Wilson, *Europe's Tragedy*, pp. 411–12.

64. Baumgardt, *Johannes Kepler: Life and Letters*, p. 163.

65. Ibid., pp. 164–65.

CHAPTER 8: THE FINAL YEARS (1626–1630)

1. Carola Baumgardt, *Johannes Kepler: Life and Letters* (New York: Philosophical Library, 1951), p. 165. There is a minor mystery here. The letter in which Kepler gives these details about his journey also refers to his "six" children, although at that stage only five were still alive. It is as if Susanna, while in Regensburg, had given birth to another child not recorded elsewhere. I am grateful to Anne Morris for our discussion on this.

2. Max Caspar, *Kepler* (New York: Dover, 1993), pp. 320–21.

3. There is a plaque on the wall of the house commemorating this.

4. The house itself no longer exists, but there is a plaque on the new building on that site informing passersby of Kepler's stay at 3 Rabengasse.

5. The information for these paragraphs comes from Ulm Museum.

6. Jonas Saur's name is recorded on the title sheet of the *Rudolphine Tables* (Johannes Kepler, *Omnia Opera*, ed. Christian Frisch, 8 vols. (Frankfurt: Heyder & Zimmer, 1858–1871), 6:613.

7. Caspar, *Kepler*, pp. 322–23.

8. Ibid., p. 324.

9. The points made here about the frontispiece are fairly self-evident. The Kepler Museum in Prague, for example, has a full explanation of them.

10. The specific point about the halo was pointed out to me by the staff at the Kepler Museum in Weil der Stadt.

11. Johannes Kepler, "Preface to the *Rudolphine Tables*," trans. Owen Gingerich and William Walderman, *Quarterly Journal of the Royal Astronomical Society* 13 (1972): 367.

12. Ibid., p. 370.

13. Ibid.

14. Caspar, *Kepler*, pp. 324–25, 331.

15. Baumgardt, *Johannes Kepler: Life and Letters*, p. 166.

16. Caspar, *Kepler*, pp. 338–39, 342–43, & 345–46.

17. Baumgardt, *Johannes Kepler: Life and Letters*, pp. 174–75.

18. Caspar, *Kepler*, p. 348.

19. Ibid., p. 350.

20. Baumgardt, *Johannes Kepler: Life and Letters*, p. 189.

21. Ibid., pp. 176–77.

22. I am grateful to Herr Wolfgang Schütz, of the Kepler Museum in Weil der Stadt, for confirming this by drawing my attention to the passage in Bernegger's letter of March 12/22, 1630, which reads "ex voto nostro in primario urbis templo Christiano," which translates as "at our request in the principal church of the town," i.e., in the cathedral. Johannes Kepler, *Gesammelte Werke*, ed. Walther von Dyck and Max Caspar, 22 vols. (Munich: C. H. Beck, 1937–2012), 18:422.

23. Baumgardt, *Johannes Kepler: Life and Letters*, pp. 184–86.

24. We know this from the letter of an otherwise unknown individual named Fischer. The letter is included in Baumgardt, *Johannes Kepler: Life and Letters*, p. 196.

25. This explanation was given to me by a member of staff at the Kepler Museum in Regensburg.

26. Caspar, *Kepler*, p. 359. The translation is my own. Neither Koestler nor (even more so) Caspar or Baumgardt seems to me to have quite got it right.

CHAPTER 9: EPILOGUE

1. From a letter by Gassendi of January 13, 1631, to Wilhelm Schickard. Quoted in Arthur Beer and Peter Beer, *Proceedings of Conferences Held in Honor of Kepler* (Oxford: Pergamon, 1975), introduction.

2. There is some dispute as to whether Bartsch died of the plague in 1632, 1633, or 1634. Max Caspar opts for 1633 (Max Caspar, *Kepler* [New York: Dover, 1993], p. 364), but the careful researches of Edward Rosen (his translation of *Somnium* [New York: Dover, 1967], p. 193) have found sources for all three of these years.

3. Information on Kepler's children was supplied by Frau Boockmann (Kepler Commission, Munich).

4. As proclaimed on the title page of *Somnium*, Johannes Kepler, *Omnia Opera*, ed. Christian Frisch, 8 vols. (Frankfurt: Heyder & Zimmer, 1858–1871), vol. 8, pt. 1, p. 27.

5. Caspar places Susanna's death in September 1638 (Caspar, *Kepler*, p. 364). Farber places it in 1636 (Beer and Beer, *Proceedings of Conferences Held in Honor of Kepler*, p. 184).

6. Recorded, for example, in J. L. E. Dreyer, *A History of Astronomy from Thales to Kepler* (New York: Dover 1953), p. 415.

7. It was Stillman Drake (Beer and Beer, *Proceedings of Conferences Held in Honor of Kepler*, p. 242) who hit the nail on the head when he pointed out that "Kepler's physics was the least satisfactory element in his astronomy, and Galileo's astronomy was the least satisfactory part of his physics."

8. Dinsmore Alter, *Pictorial Guide to the Moon* (London: Arthur Barker, 1964), pp. 4–5.

9. Johannes Kepler, *Gesammelte Werke*, ed. Walther von Dyck and Max Caspar, 22 vols. (Munich: C. H. Beck, 1937–2012), 1:24. "Mundum igitur totum figura claudi sphaerica, abundè satis disputauit ARISTOTEL . . ."

10. Strictly, final confirmation that the Earth really does move around the Sun was provided much earlier, with the discovery of the aberration of light, by Bradley in 1729.

11. See, for example, Paul Murdin and Margaret Penston, eds., *The Canopus Encyclopedia of Astronomy* (Bristol, UK: Canopus, 2004), p. 288.

12. Auguste Comte, *The Positive Philosophy* (1842), book II, chap. 1.

13. I am grateful to Dr. Keith Orrell, of the Norman Lockyer Observatory Society, for clarifying these details.

14. For a very good summary of these events, see Harry Nussbaumer and Lydia Bieri, "Who Discovered the Expanding Universe?," Cornell University Library, accessed December 14, 2014, http://xxx.lanl.gov/abs/1107.2281.

15. See Alan H. Guth, *The Inflationary Universe* (London: Vintage, 1998), where this is described in much greater detail.

16. Lawrence M. Krauss, *A Universe from Nothing* (London: Simon & Schuster, 2012).

17. Max Tegmark, *Our Mathematical Universe* (London: Allen Lane, 2014), p. 106.

18. Johannes Kepler, *Gesammelte Werke*, ed. Walther von Dyck and Max Caspar, 22 vols. (Munich: C. H. Beck, 1937–2012), 4:304: "iam erant mihi apud BRVNI innumerabilitates parata vincula et carcer, imò potiùs exilium in illo infinito."

BIBLIOGRAPHY

Alter, Dinsmore. *Pictorial Guide to the Moon*. London: Arthur Barker, 1964.

Armitage, Angus. *The World of Copernicus*. Dublin: Mentor, 1951.

Baumgardt, Carola. *Johannes Kepler: Life and Letters*. New York: Philosophical Library, 1951.

Beer, Arthur, and Peter Beer, eds. *Proceedings of Conferences Held in Honor of Kepler*. Oxford: Pergamon, 1975.

Berry, Arthur. *A Short History of Astronomy*. New York: Dover, 1961.

Bronowski, Jacob. *The Ascent of Man*. London: BBC, 1976.

Burke-Gaffney, Michael W. "Kepler and the Star of Bethlehem." *Journal of the Royal Astronomical Society of Canada*. Toronto, December 1937.

Carr, Bernard, ed. *Universe or Multiverse?* Cambridge: Cambridge University Press, 2007.

Caspar, Max. *Kepler*. New York: Dover, 1993.

Chadwick, Owen. *Penguin History of the Church—The Reformation*. London: Penguin, 1990.

Christianson, John Robert. *On Tycho's Island*. Cambridge: Cambridge University Press, 2000.

Copernicus, Nicolaus. *On the Revolutions*. Translated by Edward Rosen. Baltimore: Johns Hopkins University Press, 1978.

———. *On the Revolutions of Heavenly Spheres*. Translated by Charles Glenn Wallis. Amherst, NY: Prometheus Books, 1995.

Davies, Norman. *Europe: A History*. London: Pimlico, 1997.

Donahue, William H. *Selections from Astronomia Nova*. Green Cat Books, 2008.

Donne, John. *The Complete Poetry and Selected Prose of John Donne*. Edited by Charles M. Coffin. London: Modern Library (Random House), 2001. Kindle Edition.

Dreyer, J. L. E. *A History of Astronomy from Thales to Kepler*. New York: Dover, 1953.

———. *Tycho Brahe: A Picture of Scientific Life & Work in the Sixteenth Century*. Edinburgh: Black, 1890.

Ehrman, Bart. *The New Testament*. Oxford: Oxford University Press, 2008.

Ellerbe, Helen. *The Dark Side of Christian History*. Windermere, FL: Morningstar and Lark, 1995.

Evans, R. J. W. *Rudolph II and His World*. London: Thames and Hudson, 1997.

Ferguson, Kitty. *Tycho & Kepler*. New York: Walker, 2002.

Finkelstein, Israel, and Neil Asher Silberman. *The Bible Unearthed: Archaeolo-*

gy's New Vision of Ancient Israel and the Origin of Its Sacred Texts. New York: Touchstone, 2002.

Finocchiaro, Maurice A. *The Galileo Affair.* Berkeley: University of California Press, 1989.

Galileo. *The Sidereal Messenger.* Translated by Albert Van Helden. Chicago: University of Chicago Press, 1989.

Gingerich, Owen. *The Book Nobody Read: In Pursuit of the Revolutions of Nicolaus Copernicus.* London: Heinemann, 2004.

———. *The Eye of Heaven.* New York: American Institute of Physics, 1993.

Greene, Brian. *The Elegant Universe.* London: Jonathan Cape, 1999.

Guth, Alan H. *The Inflationary Universe.* London: Vintage, 1998.

Hawking, Stephen, and Leonard Mlodinow. *The Grand Design.* London: Bantam, 2010.

Hoskin, Michael, ed. *The Cambridge Concise History of Astronomy.* Cambridge: Cambridge University Press, 1999.

Jardine, Nicholas. *The Birth of History and Philosophy of Science.* Cambridge: Cambridge University Press, 2009.

Kepler, Johannes. *Mysterium Cosmographicum (The Secret of the Universe).* Translated by A. M. Duncan. Norwalk, Connecticut: Abaris Books, 1999.

———. *Astronomia Nova (The New Astronomy).* Translated by William H. Donahue. Cambridge: Cambridge University Press, 1992.

———. *Conversation with Galileo's Sidereal Messenger.* Translated by Edward Rosen. New York: Johnson Reprint Corporation, 1965.

———. *Epitome of Copernican Astronomy.* Translated by Charles Glenn Wallis. Amherst, NY: Prometheus Books, 1995.

———. *Gesammelte Werke.* Edited by Walther von Dyck and Max Caspar. Munich: C. H. Beck, 1937–2012.

———. *Harmonies of the World.* Translated by Charles Glenn Wallis. Amherst, NY: Prometheus Books, 1995.

———. Johannes Kepler, *Omnia Opera.* Edited by C. Frisch. Frankfurt: Heyder & Zimmer, 1858–1871.

———. *L'Étoile Nouvelle dans le Pied du Serpentaire (On the New Star).* Translated by Jean Peyroux. Paris: A. Blanchard, 1998.

———. "On Giving Astrology Sounder Foundations." Translated by J. V. Field. *Archive for History of Exact Sciences* 31 (1984): 189–272.

———. *Optics.* Translated by William H. Donahue. Santa Fe, NM: Green Lion Press, 2000.

———. "Preface to the *Rudolphine Tables,*" translated by Owen Gingerich and William Walderman. London: *Quarterly Journal of the Royal Astronomical Society* 13 (1972).

———. *Selbstzeugnisse.* Translated into German by Franz Hammer. Stuttgart: Verlag, 1971.

————. *The Six-Cornered Snowflake.* Translated by Jacques Bromberg. Philadelphia: Paul Dry, 2010.

————. *Somnium.* Translated by John Lear and Patricia Kirkwood, as *Kepler's Dream.* Berkeley: University of California Press, 1965.

————. *Somnium.* Translated by Edward Rosen. New York: Dover, 1967.

————. *Somnium.* Kindle Edition. Nexum Ediciones, 2013.

Koestler, Arthur. *The Sleepwalkers.* London: Penguin, 1964.

————. *The Watershed: A Biography of Johannes Kepler.* London: Heinemann, 1960.

Koyré, Alexandre. *The Astronomical Revolution.* London: Routledge, 2009.

————. *From the Closed World to the Infinite Universe.* London: Forgotten Books, 2008.

Krauss, Lawrence M. *A Universe from Nothing.* London: Simon & Schuster, 2012.

Kuhn, Thomas. *The Copernican Revolution.* Cambridge, MA: Harvard University Press, 1957.

Machamer, Peter, ed. *The Cambridge Companion to Galileo.* Cambridge: Cambridge University Press, 1998.

McBride, Neil, and Iain Gilmour, eds. *An Introduction to the Solar System.* Cambridge: Cambridge University Press, 2004.

Mitton, Simon, ed. *The Cambridge Encyclopedia of Astronomy.* London: Jonathan Cape, 1977

Moore, Patrick. *The Solar System.* London: Methuen, 1958.

Morison, Ian. *A Journey through the Universe.* Cambridge: Cambridge University Press, 2014.

Murdin, Paul, and Margaret Penston, eds. *The Canopus Encyclopedia of Astronomy.* Bristol, UK: Canopus, 2004.

Parker, Geoffrey. *Europe in Crisis, 1598–1648.* Glasgow: Fontana, 1979.

Rees, Martin J. *Before the Beginning: Our Universe and Others.* London, Simon & Schuster, 1997.

Rosen, Edward, trans. *Three Copernican Treatises.* New York: Dover, 1959.

————. *Copernicus and his Successors.* London: Hambledon, 1995.

Russell, Bertrand. *History of Western Philosophy.* London: Unwin 1969.

Sanders, E. P. *The Historical Figure of Jesus.* London: Penguin, 1995.

Schütz, Wolfgang. *Kepler und die Nachwelt.* Weil der Stadt, Germany: Scharpf, 2009.

Shirley, John W. *Thomas Harriot: A Biography.* Oxford: Clarendon, 1983.

Smart, William M. *Spherical Astronomy.* Cambridge: Cambridge University Press, 1965.

Sobel, Dava. *Galileo's Daughter: A Historical Memoir of Science, Faith, and Love.* London: Fourth Estate, 1999.

————. *A More Perfect Heaven: How Copernicus Revolutionized the Cosmos.* London: Bloomsbury, 2011.

Stephenson, Bruce. *Kepler's Physical Astronomy*. Princeton, NJ: Princeton University Press, 1987.

Tegmark, Max. *Our Mathematical Universe*. London: Allen Lane, 2014.

Thiel, Rudolph. *And There Was Light: The Discovery of the Universe*. London: André Deutsch, 1958.

Voelkel, James R. *The Composition of Kepler's* Astronomia Nova. Princeton, NJ: Princeton University Press, 2001.

Vermes, Geza. *The Nativity*. London: Penguin, 2006.

Vilenkin, Alex. *Many Worlds in One: The Search for Other Universes*. New York: Hill and Wang, 2006.

Ward, Peter, and Donald Brownlee. *Rare Earth: Why Complex Life Is Uncommon in the Universe*. New York: Springer-Verlag, 2000.

White, Andrew. *History of the Warfare of Science with Theology in Christendom*. Amherst, NY: Prometheus Books, 1993.

White, Michael. *Galileo Antichrist: A Biography*. London: Weidenfeld & Nicolson, 2007.

Wilson, Peter H. *Europe's Tragedy*. London: Penguin, 2010.

Wootton, David. *Galileo: Watcher of the Skies*. New Haven, CT, and London: Yale University Press, 2010.

INDEX

Boldfaced numbers refer to images and figures

THE KEPLER MISSION TO DISCOVER EXTRASOLAR PLANETS. An artist's impression of the Kepler Mission, which, at the time of publication, had discovered well over 4,000 planetary candidates orbiting other stars. Over 1,000 of these have so far been confirmed as planets. *Credit: NASA/Kepler Mission/Wendy Stenzel.*